Radiation chemistry

AN INTRODUCTION

Radiation chemistry

AN INTRODUCTION

A. J. Swallow
Ph.D., D.Sc., Sc.D., F.R.I.C.

Longman

LONGMAN GROUP LIMITED
London

Associated companies, branches and representatives throughout the world

Published in the U.S.A. by Halsted Press,
a Division of John Wiley and Sons, Inc.

First published 1973

ISBN 0 582 46286 X

Printed in Great Britain by
William Clowes & Sons, Limited, London, Beccles and Colchester

Preface

Knowledge of the chemical effects produced by high-energy radiations is beginning to be sufficiently well founded for radiation chemistry to be regarded as a distinct scientific discipline. The subject is dealt with briefly in textbooks of physical chemistry, and in greater detail in an ever-increasing number of other publications ranging from simple guides through textbooks, specialist monographs and reviews to original scientific papers. The present book aims to give a balanced picture of the subject as a whole. It has been written not just for those specializing in radiation chemistry, but also for those whose main interests are in related fields such as nuclear technology, radiation physics, radiochemistry, photochemistry, chemical kinetics, polymer science and radiation biology.

This book attempts to get down to fundamentals by discussing typical work. It does not attempt to discuss the subject exhaustively, and although reference is made to some hundreds of papers, there is no reference to thousands of others which have contributed directly and significantly to the development of the subject. So that there should be no impression that all the best work has been cited, the names of scientists have, with few exceptions, been excluded from the text, and there is no author index. It is stressed that radiation chemistry is still actively developing. Indeed each month sees the appearance of several hundred new papers. Consequently those wishing to keep fully up-to-date will certainly need to consult the latest journals and reviews.

This book has grown out of countless lectures, and has benefited from the author's participation in numerous conferences and his visits to laboratories in many parts of the world. The author is grateful to bodies such as the Royal Society, the British Council, the Wellcome Trust, the United States and other Atomic Energy Commissions, the International Atomic Energy Agency, and Universities, Hospitals and Industry throughout the world for making such visits possible. He is most grateful to his colleagues at the Paterson Laboratories for a stimulating environment and for much help and advice, and also thanks many other scientists for sending copies of their work and for helpful discussions. He is especially grateful to those from other laboratories who have read parts of this book in manuscript and in particular thanks Drs A. O. Allen, J. H. Baxendale, A. Charlesby, G. R. A. Johnson, W. J. Meredith, R. S. Nelson, R. L. Platzman, M. A. J. Rodgers, C. Schneider and G. Stein. He is greatly indebted to Dr B. J. Parsons for help in checking all the problems and in correcting the proofs.

Contents

Acknowledgements

Thanks are due to the following for supplying material for the preparation of illustrations:

Academic Press, London and New York (Figs. 3.3 and 4.9); Argonne National Laboratory and Dr J. H. Kittel (Fig. 5.4); Atomic Energy of Canada Ltd., Commercial Products (Fig. 1.2); Atomic Energy Research Establishment and Professor A. Charlesby (Fig. 9.3); Professor A. Charlesby (Fig. 9.5); High Voltage Engineering Corporation, Massachusetts (Figs. 1.4 and 1.7); McGraw-Hill Book Company and M. Lewis (Fig. 2.7); Pergamon Press (Fig. 1.5); Raychem Ltd. (Fig. 9.6).

Sources of radiation

Some years after the discovery of cathode rays, but soon after the discoveries of X-rays and of radioactivity, scientists began to enquire into the chemical effects which could be produced by high-energy radiations. They have now gained a great deal of insight into the fundamental processes occurring. They have discovered many new chemical species. Some of the findings are of great interest in themselves and also important for other branches of pure and applied science.

The development of radiation chemistry has been associated with discoveries of natural radiations and with inventions of new ways of producing artificial radiation. Indeed the modern subject may be said to have originated with the development of atomic energy, for which it was essential to learn how to minimize any harmful effects associated with reactors. This chapter introduces radiation chemistry by discussing briefly the various sources of high-energy radiation. At the same time it indicates the kinds of chemical change to which irradiation can give rise.

Natural radioactivity

In 1896, just after Roentgen's discovery of X-rays, Henri Becquerel found that the element uranium, whatever its chemical form, and without any kind of activation, appeared to emit a penetrating form of radiation which could be detected by the blackening of a photographic plate. Following the discovery of radium by the Curies, Becquerel observed a number of interesting effects of a chemical and biological nature produced by the radiation and in 1901 published one of the first papers on radiation chemistry as such [1].

The most striking radiation-chemical effect produced by radium is decomposition of water. Also glass apparatus becomes discoloured and biological effects can be produced. The biological effects can be harmful, but they can be put to good use in the treatment of cancer. It is now clear that such effects are not confined to radium: experiments with radioactive substances of high specific activity are always complicated by chemical effects due to the radiation, and stringent safety precautions have to be maintained because of the possibility of danger to health. Many types of radiation are now used in cancer treatment.

Radium itself continued to be used as a source of γ-rays for experimental purposes as well as for radiotherapy until after the Second World War when it became superseded by the much cheaper artificial radioisotopes, especially cobalt-60. The radium was often used in the form of dry radium sulphate enclosed in a sealed container to prevent the emission of α- and β-particles and to eliminate the spread of gaseous decay products. Polonium-210 (RaF) used to serve, and indeed still serves, as a useful laboratory source of α-particles (energy 5.30 MeV).† Radon in equilibrium with its decay products is another α-particle emitter, giving α-particles of energy 7.7, 6.0 and 5.5 MeV together with some β-radiation. Much of the pioneering work in radiation chemistry was done by mixing radon with gaseous reactants.

The largest terrestrial sources of radiation are potassium-40, uranium-238, uranium-235 and thorium-232. The total energy emission from the potassium-40 in the Earth's crust may be estimated to be about 4×10^{12} W, which is somewhat greater than the likely emission from waste fission products from nuclear power reactors of the future. Were it not for the heat generated by radioactivity, the Earth would have cooled down much more rapidly than it has. Moreover before radiation energy is finally degraded to heat it must produce physical and chemical changes, and some of these may be significant for the history of the Earth. It is possible, for example, that radiations of radioactive origin may have been partly or entirely responsible for the formation of the organic compounds from which life developed [2]. It is also possible that the effect of radiation on hydrocarbons could have been responsible for the generation or the modification of petroleum in certain deposits. Another effect of natural radioactivity is to introduce radiation damage into certain minerals. This can be released in the form of heat or light by heating (pp. 99, 112).

† 5.30 million electron volts. An electron volt is the energy acquired by an electron in falling through a potential difference of one volt, that is, 1.602×10^{-19} J.

Nuclear reactors

The construction of the first nuclear reactor in Chicago in 1942 initiated a new phase in radiation science. Nuclear reactors are powerful sources of fast and slow neutrons and γ-rays, and their radiations affect the various materials of which the reactor is constructed as well as being useful for experimental purposes. A diagram of a typical 'swimming pool' reactor is shown in Fig. 1.1 [3].

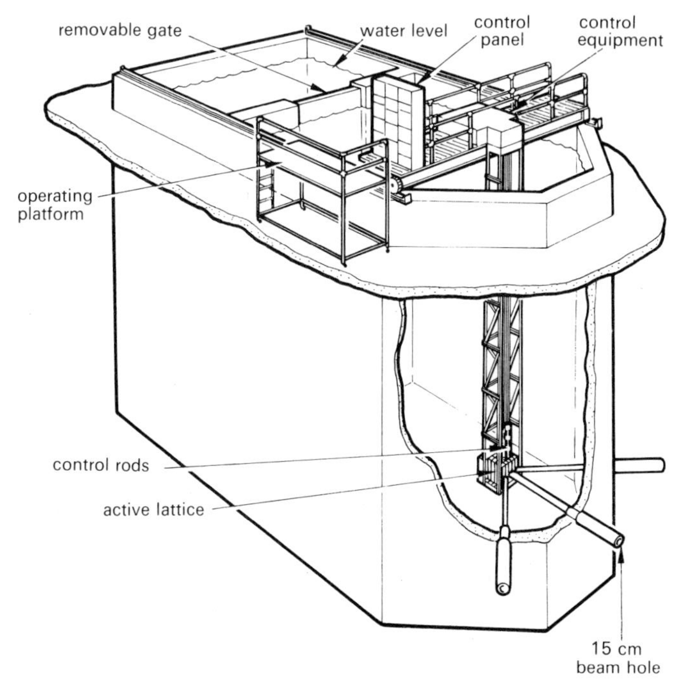

Fig. 1.1 Research reactor of the widely used 'swimming pool' type.

The energy of the fission process makes its appearance in the first instance in the form of various kinds of radiation energy as shown in Table 1.1. The largest proportion appears in the form of kinetic energy of the fission fragments. The fast fragments are a form of radiation resembling α-particles in that they are energetic nuclei of atoms. Their energy is rapidly degraded to heat, but during the slowing down process they may cause radiation damage to the fuel element (see p. 105). In principle it might be possible to utilize the high energy of the fission fragments in ways which are more subtle than by simply allowing heat to be produced.

One way to do this would be to construct a special kind of nuclear reactor with the fuel in, for example, a gaseous or finely divided form so that the fragments could escape into the surrounding medium and change it chemically. Air could be irradiated in this way for example, giving oxides of nitrogen which could be used to give nitric acid [4]. The idea does not appear to have been used commercially however, although a great deal of development work has been done.

Table 1.1 *Percentage of the energy of the fission reaction which appears in various forms*

Kinetic energy of fission fragments	81
Kinetic energy of fission neutrons	2.5
Energy of fission γ-rays	2.5
Energy of γ-rays produced by neutron capture in constructional and other materials together with energy from decay of radioactive products formed	2.5
Energy of γ-rays from decay of fission products	3
Energy of β-particles from decay of fission products	3.5
Kinetic energy of neutrinos	5
	100

All the other radiations present in a nuclear reactor, except for the neutrinos, which do not interact appreciably with matter, can also cause radiation effects. The effects are often damaging to the materials of the reactor, and such factors must be taken into account in design, especially in the construction of large and powerful reactors. The greatest effects occur on substances which contain labile chemical bonds, such as plastics, organic compounds, water or carbon dioxide, but there are also highly important effects with materials like metals, graphite and ceramics (Chapter 5).

Nuclear reactors are useful for the deliberate irradiation of materials. They are not the best available sources of β- or γ-radiation as such because in a reactor the neutrons often cause the material to become radioactive, and this complicates handling. Also the exact dose of radiation received cannot easily be estimated since there are many variations in radiation intensity during the routine operation of a reactor. Reactors are, however, excellent sources of fast neutrons. Reactors are especially valuable for experiments on inorganic solids where the contribution to the radiation effect due to β- and γ-rays is often small (p. 98).

Artificial radioactivity

Radioisotopes were first produced artificially in 1934, but it was only when they began to be produced in large quantities in nuclear reactors that their radiation-chemical effects had to be considered seriously. They are made by inserting suitable elements into the reactor. Certain radioisotopes are produced as fission products. The radiations from fission products cause damage to the materials that are used in processing used reactor fuel, which include water, complexing agents, organic solvents and ion-exchange resins. After separation the fission products must be stored in places where they cannot harm life and under conditions such that the radiations could not cause damage which would enable the fission products to escape into the environment.

The potential for the industrial production of radioisotopes is considerable. Moreover there are problems in disposing of the large quantities of fission products associated with the growth of nuclear power. For these reasons there has been a drive to discover practical uses for radiation. One use for small amounts of certain radioisotopes is as radioactive tracers. Radiation damage can be a consideration in the use of radioactive tracers, and is serious in the prolonged storage of labelled organic compounds [5]. The nuclear transformation accompanying decay makes negligible contribution to the damage since few molecules are affected, and in any case the molecules damaged by this process are in general no longer labelled, but self-absorption of radiation can cause appreciable change in the labelled material.

Cobalt-60, made by reactor irradiation of cobalt metal, is widely used as a source of high-energy radiation for experimental purposes. Cobalt-60 emits two γ-rays, 1.17 and 1.33 MeV per disintegration, and has a half-life of 5.27 years. For use as a radiation source the metal is encapsulated in, for example, stainless steel to eliminate the possibility of spreading the activity. Often the source is in the form of a hollow cylinder or of something approximating to a hollow cylinder such as a cylindrical array of rods. This simplifies the irradiation procedure because the intensity of radiation within a cylinder is approximately independent of position. Furthermore large samples can be given a dose of radiation which is uniform throughout their volume. A compact irradiation unit, shielded by lead, is shown in Fig. 1.2. Another common arrangement is to use a room with thick concrete walls. The operator enters the room while the source is withdrawn to a safe position, and sets up his apparatus. He then leaves the room and moves the source into position by remote control. The fission product caesium-137 (decaying to ^{137}Ba with 92 per cent of the disintegrations resulting in a γ-ray of energy 0.66 MeV, half-life 30 years) is also used as an experimental radiation source.

Several applications of radiation have now found their way into industrial practice. These include sterilization of medical supplies as well as promotion of chemical reaction. A diagram of one of the first pilot-plant

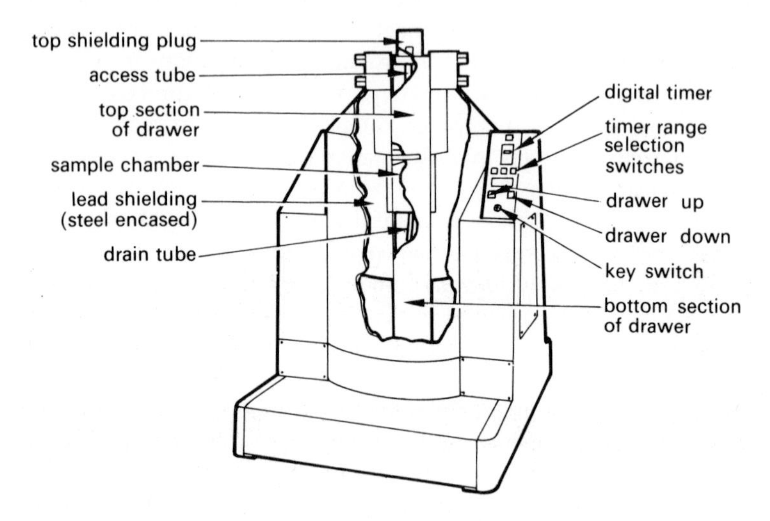

Fig. 1.2 'Gammacell 220' cobalt-60 irradiation unit (Atomic Energy of Canada). The sample chamber, which is about 15 cm in diameter, is surrounded by a fixed source cage. The chamber forms part of the vertical drawer which is moved upwards to bring the chamber outside the shielding for loading and unloading.

irradiation units (150 000 curies† of cobalt-60) is shown in Fig. 1.3 [6]. Fission products could be used if they cost less to prepare in an acceptable form. Other sources of radiation which have been considered for industrial purposes include used fuel elements whose activity is decaying to a level where they can be handled in separation plant, and fluids such as liquid sodium or indium sulphate solutions, which can be circulated through a reactor to give radioactivity of short half-life, the samples being subjected to irradiation outside the reactor while the decaying fluid returns to the reactor for reactivation.

A useful β-particle source is tritium (maximum β-ray energy 0.0186

† The curie (Ci) is a unit of radioactivity defined as the quantity of any radioactive nuclide in which the number of disintegrations per second is 3.700×10^{10}.

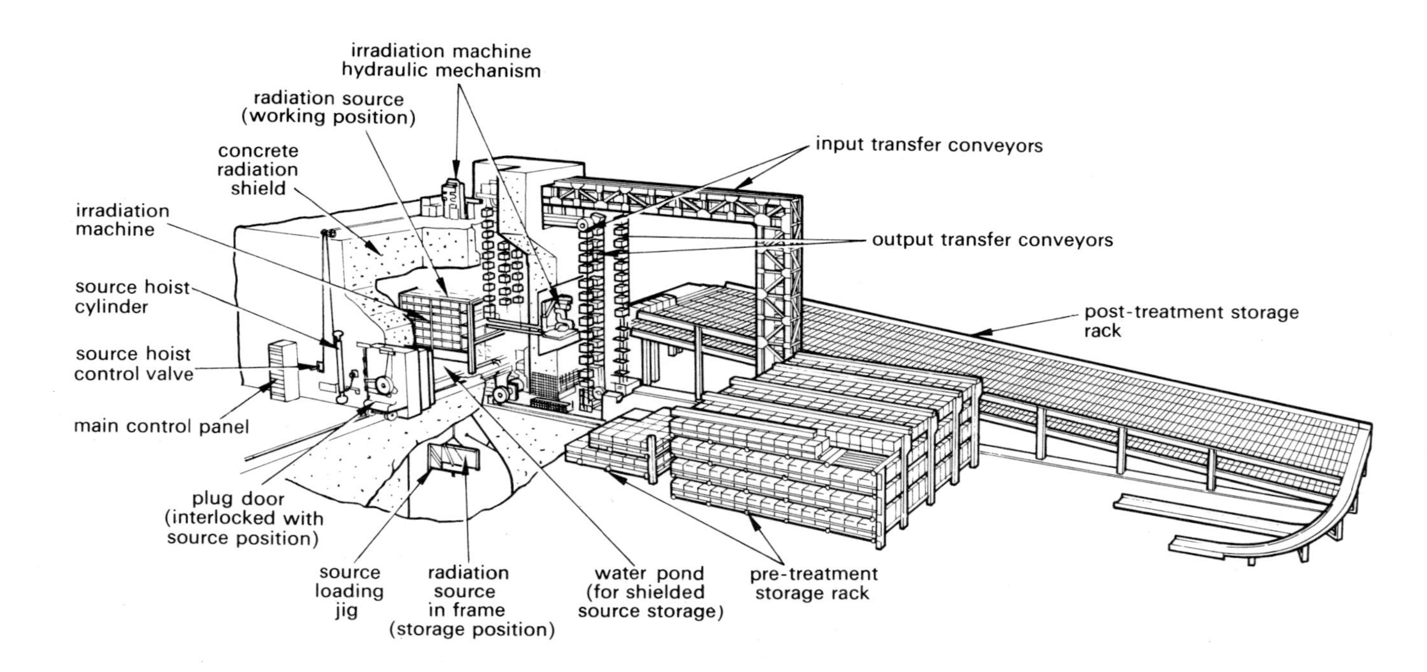

Fig. 1.3 Unit for experimental γ-irradiations on a pilot-plant scale.

MeV, mean energy 0.005 65 MeV, half-life 12.26 years)† which in the form of tritium gas, tritiated water or other substances can be mixed with a chemical whose radiolysis it is desired to study. This method has also been applied to the synthesis of tritium-labelled organic compounds, tritium atoms becoming incorporated into the organic molecules under the influence of β-rays from other tritium atoms (see p. 120).

Machine sources

The development of electrical machines to produce radiation pre-dated the discovery of radioactivity, and has since played a vital role in the progress of nuclear science. Electrical machines can produce radiations which are very similar to nuclear radiations. X-rays themselves, discovered by Roentgen, are the extranuclear radiations corresponding to γ-radiation, which may be defined as the electromagnetic radiation emitted by nuclei. There are two types of X-ray. One type is called 'bremsstrahlung'‡ and is produced when fast electrons are decelerated in the field of the nucleus. This type is responsible for the continuous X-ray spectrum in X-ray sets. The other type is the 'characteristic' X-radiation, and is produced when an electron falls from one atomic shell to another. Helium ions may be accelerated by machines, and are indistinguishable from α-rays. Artificially accelerated electrons can have energies which are comparable to (or much greater than) those of β-rays, but the correspondence is not exact because the electrically accelerated electrons do not have the same spread of energies as β-rays.

Most machines for producing radiation work on the same principle. A hot filament is used to produce electrons, or an electrical discharge is used to produce positive ions such as helium, deuterium or hydrogen ions. These are then accelerated electrically through a vacuum towards a target. It is possible to insert samples into the vacuum system for irradiation but it is more convenient to take the particle beam out of the accelerator through a thin window, and irradiate externally. X-rays can be generated using any electron accelerator. The electrons are simply allowed to fall on a target of high atomic number. This procedure gives X-rays (bremsstrahlung) possessing all energies up to that of the bombarding electrons. The same principle is used in conventional low voltage

† β-particles are emitted from nuclei with all energies up to a certain maximum. It is the maximum energy which is quoted in charts or tables of nuclear data. The mean energy is about one-third of the maximum energy. In the case of α-particles and γ-rays the emission is monochromatic and the energy quoted is the actual energy of the particle or photon.

‡ German for 'slowing-down radiation'.

X-ray sets. In many X-ray sets a steady potential is used to accelerate the electrons, but it is also possible to apply an alternating potential between electron filament and target. The electrons flow only when the target is the cathode, so that this method produces a rapidly fluctuating X-ray beam. The energy of the X-rays is then characterized by the peak voltage applied, for example, if X-rays are produced by applying a 220 kV alternating potential between filament and target, then the radiation is called 220 kVp X-radiation. X-rays find a major use in radiotherapy, and have been used for many of the pioneering experiments in radiation chemistry.

Neutrons can be generated with machine-produced radiations such as deuterons or energetic X-rays by making use of suitable nuclear reactions. Typical reactions† are $^{9}Be\ (^{2}H, n)\ ^{10}B$; $^{7}Li\ (^{2}H, n)\ ^{8}Be$; $^{2}H\ (^{2}H, n)\ ^{3}He$; $^{2}H\ (\gamma, n)\ ^{1}H$.

Besides being useful for radiotherapy and fundamental research, machines are excellent for producing radiation for industrial purposes, and machine-produced fast electrons have been used on a large scale since the mid-nineteen fifties for such purposes as the sterilization of pharmaceuticals and medical products, irradiation of polyethylene electrical components and production of shrinkable packaging film.

Van de Graaff accelerator This type of machine, first operated in 1929, can be used to accelerate electrons or positive ions. A diagram of a typical electron accelerator is shown in Fig. 1.4. Negative electricity is sprayed on a moving insulating belt and is taken by the belt to a discharge point which transfers it to the outside of a metal hemisphere. The metal hemisphere is insulated from the outer tank by a high pressure of gas. The high potential difference between the hemisphere and the lower end of the accelerator is used to accelerate electrons through the evacuated accelerating tube. The beam of electrons can be scanned magnetically so as to emerge through the window as a broad beam suitable for the irradiation of large objects.

Van de Graaff generators can produce energies of up to several MeV, but an upper limit is set by insulation difficulties. Beam currents can be up to several milliamps.

Cyclotron The conventional cyclotron, first operated in 1930, is an excellent machine for producing intense beams of positive ions with energies of up to tens of MeV. A diagram of a conventional cyclotron is shown in Fig. 1.5. Positive ions are produced in the gap between two flat semicircular copper boxes called 'dees'. The ions are accelerated towards

† Strictly the last of these reactions is not produced by γ-rays but by X-rays.

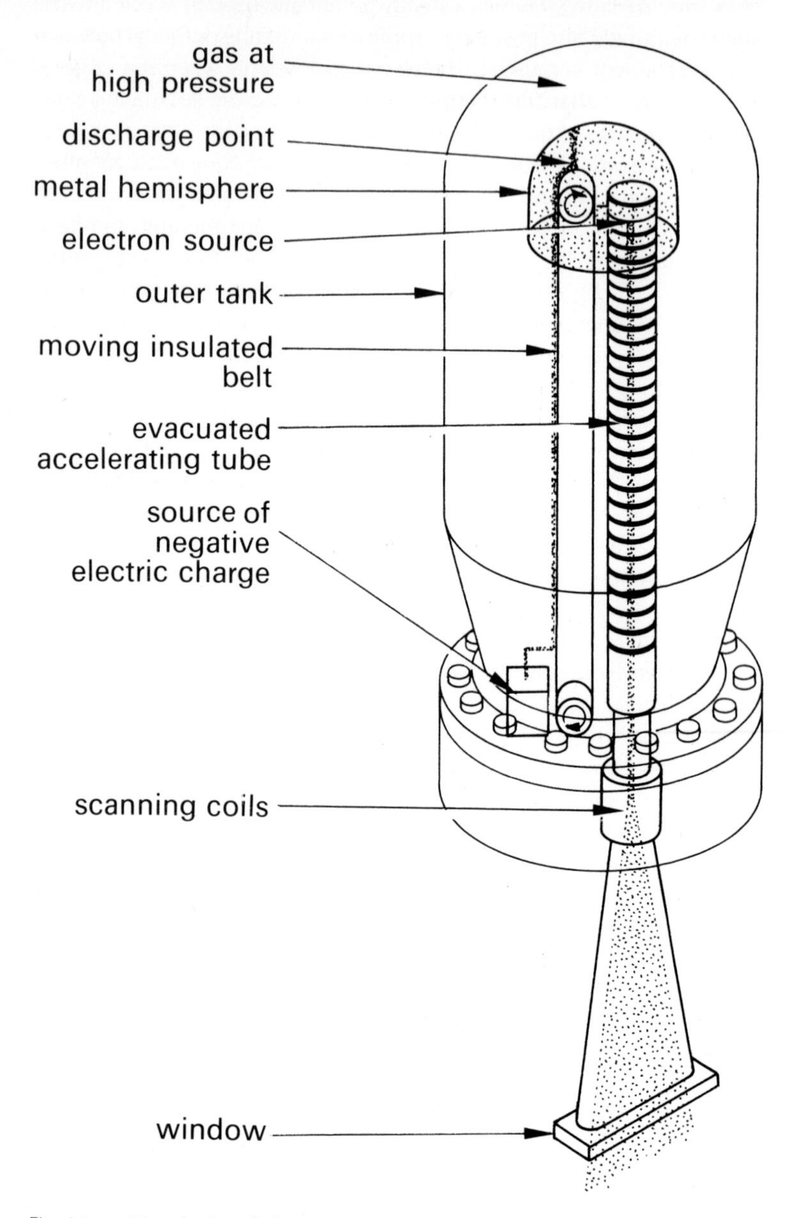

Fig. 1.4 Van de Graaff electron accelerator.

one of the dees by an electric field and are caused to pursue a semicircular path within the dee by an electromagnet situated above and below the dees. On returning to the gap the ions are again accelerated across it by a reversed field and again curve round within the dee, this time with a slightly greater radius because of their greater velocity. This process continues to take place, actuated by a high-frequency potential

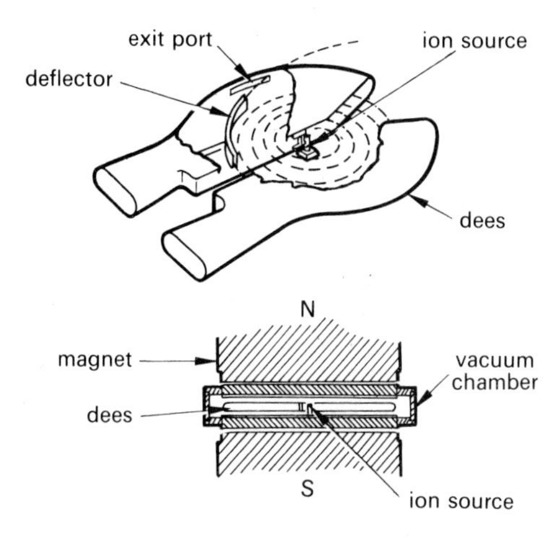

Fig. 1.5 Cyclotron.

between the dees, until the ions have spiralled to a negatively charged deflector plate which directs them outside the dees through an exit port. The beam then leaves the cyclotron through a thin window in the vacuum chamber.

The energy attainable with a conventional cyclotron is limited by the relativistic increase in mass of the particle as its energy increases. Energies of hundreds of MeV are attainable with a modified version of the cyclotron called the synchrocyclotron or frequency-modulated cyclotron. Various versions of the cyclotron have proved invaluable for investigations into the effects of densely ionizing heavy particles, a topic which has played an essential part in the advance of nuclear technology.

Resonant transformer This machine was first developed during 1937–39 for producing X-rays for the treatment of cancer. It has been extensively used since 1948 as a source of fast electrons for industrial processing.

A diagram of a 1 MeV resonant transformer is shown in Fig. 1.6 [7]. Flat pancake secondary coils are stacked above a primary winding. They are tuned to a harmonic of the frequency of alternating current used to feed the primary. Electrons are generated at the top of the permanently evacuated accelerating tube and, after being accelerated and focused,

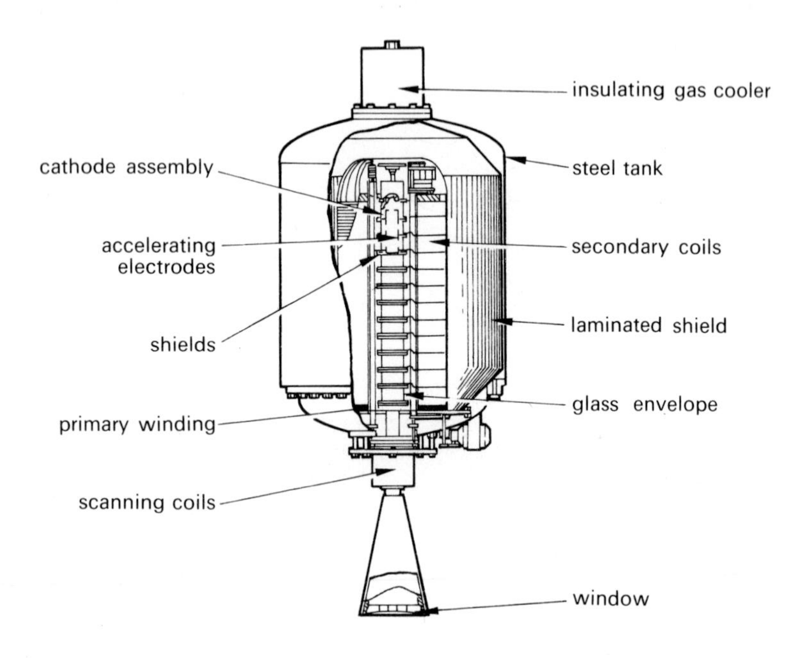

Fig. 1.6 Resonant transformer for use as a source of fast electrons.

the beam emerges through a thin steel window into the air. It may be scanned magnetically if desired. The whole unit is mounted inside a pressurized steel tank for insulation, and this is shielded with overlapping silicon iron strips to prevent loss of energy through eddy currents.

Microwave linear accelerator These machines, dating from the early nineteen fifties, can be used to accelerate electrons to almost unlimited energies. A diagram of a typical linear accelerator is shown in Fig. 1.7. A magnetron or klystron power generator is used to generate travelling waves of high-frequency microwave power. These are directed down an evacuated waveguide. Pulses of electrons are introduced into the wave-guide and become accelerated by the wave to high energies. The emer-

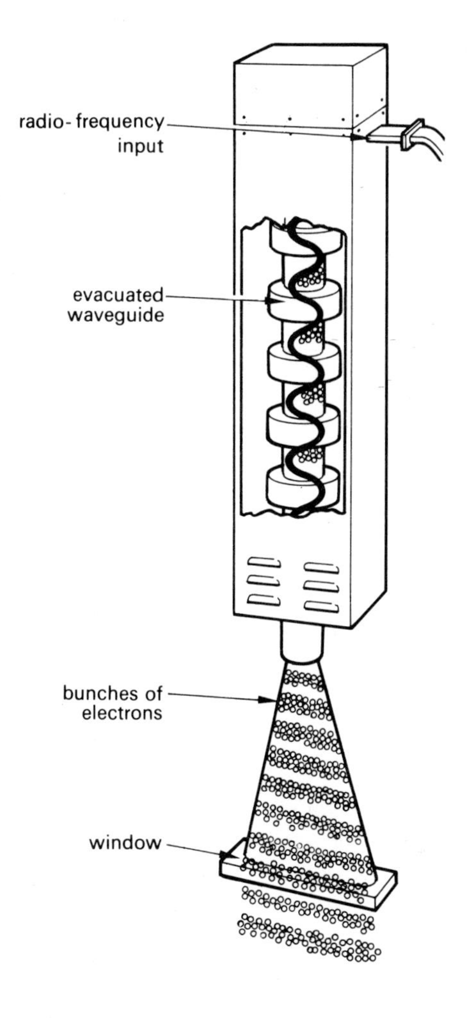

Fig. 1.7 Linear accelerator.

gent electron beam is focused and may be scanned magnetically if it is desired to have a broad beam suitable for the irradiation of large objects.

A special feature of this accelerator is that the electron beam is not continuous but takes the form of a succession of pulses, each lasting about a microsecond or less. Each pulse delivers considerable energy to the system irradiated. These features permit certain unique experiments to be done. In particular, a single pulse of radiation can be given to a system, and the transient species formed can be examined spectroscopically as first done in the flash photolysis technique in photochemistry. This technique has been outstandingly successful, and other types of accelerator have now been modified so as to permit single pulses of radiation to be given for 'pulse radiolysis' (see p. 90).

Dynamitron This machine was first introduced in 1959 and produces very high outputs of radiation energy (up to 30 kW). It contains an evacuated accelerating tube surrounded by a high voltage generator consisting of a cascaded rectifier system in which all the rectifiers are driven in parallel from a high frequency oscillator. The electron beam may be scanned and escapes into the air in the usual manner. The dynamitron principle can also be used for the acceleration of positive particles.

Pulsed field emission machine In contrast to other accelerators, in these machines, introduced in the nineteen sixties, the electrons are produced at a field emission cathode. The accelerating voltage is produced by charging a number of condensers in parallel and then discharging them in series. The machine produces radiation in pulses of great intensity but cannot easily produce energies greater than a few MeV. It is useful for pulse radiolysis, especially of gases.

Other accelerators Many accelerators besides those mentioned have been used for various aspects of radiation chemistry and indeed some have been specially developed for chemical purposes. During the nineteen sixties, for example, an important need arose out of the possibility of using radiation on an industrial scale for the curing of organic coatings on wood, metal and other materials. The technical feasibility of the process was demonstrated using high-energy electron accelerators, but it would obviously be wasteful to use high-energy electrons when only thin layers were to be treated. Special accelerators were therefore developed for the economical production of electrons with energies of 100–300 keV. New types of accelerator will doubtless continue to be developed to exploit new technology and to meet new needs.

Radiation in space [8]

Full consciousness of the radiations in space has begun to be acquired in the past few years with the aid of the space projects of the United States and the Soviet Union. The principal kinds of high-energy radiation in the astronomical environment of the Earth are (a) cosmic rays of non-solar origin (galactic cosmic rays), (*b*) the geomagnetically trapped corpuscular radiations discovered by Van Allen and his co-workers and (*c*) particles arising from solar flares (that is, cosmic rays of solar origin). High altitude nuclear bursts have also produced significant radiation in the past.

The galactic cosmic radiation consists mainly of very energetic and therefore penetrating protons and heavier ions, but the intensity is very small. In space the dose-rate is less than 20 rads per year† and our atmosphere reduces the dose-rate to 0.03 rad per year. The Van Allen radiation is more intense; it consists mainly of electrons with energies of tens of keV and protons with energies of tens or hundreds of MeV, occupying a distorted toroid lying in the equatorial plane of the Earth. The electron flux is greatest about 16 000 km away from the Earth and the proton flux about 3 500 km.

The fast electrons in the Van Allen belt can very easily be eliminated by shielding, so that radiation damage due to this source is very slight except at surfaces of objects. A shield with a thickness of 0.4 g per cm^2 will stop the electrons completely. Inside such a shield the dose-rate due to the protons in the heart of the proton belt is about 100 rads per hour. A shield of 1 g per cm^2 reduces this dose-rate by less than a half. A dose of 500–1000 rads would kill a man, so that in manned space vehicles the time which could be spent in the Van Allen belt is strictly limited. Semiconductor devices may be seriously damaged by doses as low as 100 rads.

Solar flares occur only sporadically, with the probability of a flare varying in an eleven-year cycle from about one a month to about one or two a year. Each event lasts for many hours. The intensity of the proton and α-particle emission associated with solar flares can be very great. For example, the 24 hour flare of 14–15 July 1959 (one of the more intense recorded) would have delivered about 600 rads to matter below a shielding thickness of 1 g per cm^2. The particles have energies of up to several hundreds of MeV but even so additional shielding can reduce the dose-rate several times. Since there is no escape from a solar flare once it starts, flares present one of the most serious radiation hazards to space vehicles and, apart from addition of extra shielding, the best safeguard is to avoid the periods when solar activity is at a maximum. Solar flares also present a potential danger to those flying in supersonic aircraft at high altitudes.

Artificial electron bombardment from the nuclear burst of 9 July 1962 gave dose-rates through 1.3 g per cm^2 of shielding as high as 3 rads per hour. The apparent half-life for disappearance of the radiation was initially 15 days and lengthened during the months following the burst.

† One rad equals 100 ergs per gram or 10^{-5} J per gram (see Chapter 3).

As well as true high-energy radiation, some of the ultraviolet radiation in space may be considered as high-energy radiation in that it has sufficient energy to cause ionization.† The energetic ultraviolet light is absorbed in the oxygen in the upper layers of the atmosphere, giving ozone. At one time the Earth's atmosphere consisted of water, hydrogen, methane, ammonia and nitrogen, and one theory holds that it was the absorption of energetic ultraviolet radiation in these substances which produced the simple organic chemicals which were necessary precursors to life.

PROBLEMS

1. A sample of ^{14}C-labelled sucrose has a specific activity of 50 mCi per gram. Calculate the percentage of molecules decomposed after storage for one year if half the β-radiation emitted (maximum energy = 0.16 MeV) is absorbed in the sample and decomposes sucrose at the rate of 5 molecules per 100 eV.

[Ans: 4.4]

2. 10 grams of copper are bombarded for one second with 2 MeV electrons at a beam current of 100 μA. Assuming all the electron energy enters the sample and is converted into heat, calculate the energy absorbed in (*a*) watt-hours, (*b*) electron volts. (*c*) What is the temperature rise of the sample? (specific heat of copper = 0.39 Jg^{-1}).

[Ans: (*a*) 0.056 (*b*) 1.26×10^{21} (*c*) 52°C]

3. What quantity of polonium-210 would be needed to produce the same output of α-particle energy as a cyclotron producing 100 μA of 40 MeV α-particles?

[Ans: 6.4×10^4 curies]

4. (*a*) Calculate the cost of radiation per kWh from a 10 kW accelerator costing $70 000 and running for 4 000 hours per year if the machine must be depreciated to zero (linearly) over 5 years and all other costs including maintenance, salaries, overheads and so on cost $40 000 per year. (*b*) If the machine can produce a chemical of molecular weight 100 at the rate of 3 molecules per 100 eV, calculate the annual output in kg and the cost of the radiation treatment in $ per kg.

[Ans: (*a*) $1.35 (*b*) 4 480, 12.05]

5. (*a*) What is the output of γ-ray energy in eV per second of a cobalt-60

† The energy of a photon is equal to Planck's constant *h* multiplied by the frequency. Frequency equals velocity divided by wavelength.

source of 10 kCi? (*b*) What would be the output 5.3 years later? (*c*) What power in watts of an ultraviolet lamp would be needed to produce the same output of energy at a wavelength of 366 nm as the total energy output from a 10 kCi ^{60}Co source if the lamp emits 5 per cent of its energy at 366 nm? (*d*) How many quanta at 366 nm would the lamp be emitting per second?

[Ans: (*a*) 9.25×10^{20} (*b*) 4.62×10^{20} (*c*) 2970 (*d*) 2.73×10^{20}]

REFERENCES

1. H. Becquerel. *Compt. rend.*, **133**, 709–12 (1901), 'Sur quelques effets chimiques produits par le rayonnement du radium'.
2. M. Calvin. *Chemical Evolution*. Clarendon Press, Oxford, 1969.
3. W. M. Breazeale, R. G. Cochran and K. O. Donelian. *U.N. Intern. Conf. Peaceful Uses Atomic Energy 1st Geneva*, 1955, **2**, 420–27, 'The swimming pool reactor and its modifications'.
4. P. Harteck and S. Dondes. *Nucleonics*, **14** (7), 22–5 (1956), 'Producing chemicals with reactor radiations'.
5. R. J. Bayly and E. A. Evans. 'Storage and Stability of Compounds Labelled with Radioisotopes', *Review 7*. The Radiochemical Centre, Amersham, 1968.
6. S. Jefferson, G. S. Murray and F. Rogers, in *Large Radiation Sources in Industry*. IAEA, Vienna, 1960, **1**, pp. 179–85, 'Gamma radiation sources of the technological irradiation group, U.K.A.E.A.'
7. J. A. Knowlton, G. R. Mahn and J. W. Ranftl. *Nucleonics*, **11** (11), 64–6 (1953), 'The resonant transformer: a source of high-energy electrons'.
8. J. W. Haffner. *Radiation and Shielding in Space*. Academic Press, New York, N.Y., 1967.

Absorption of radiation in matter

The Grotthuss-Draper law of photochemistry states that radiation has to be absorbed by a system if it is to produce chemical change. With only slight modification the law applies to radiation chemistry as well as to photochemistry although, because of the greater variety of radiations and their higher energy, a greater number of absorption processes has to be considered.

The study of the processes by which high-energy radiations are absorbed in matter forms the subject of radiation physics, many of the principles of which were established long before the current interest in radiation chemistry began [1–3]. Textbooks of nuclear or radiation physics should be consulted for fuller details than can be given here.

X- and γ-rays

There are three important processes by which X- or γ-rays can give energy to matter: photoelectric absorption, Compton scattering and pair production. In contrast to the processes by which less energetic radiations give energy to matter, the processes are entirely independent of chemical binding.

Photoelectric absorption This process was first established for light in the early part of the twentieth century. In the photoelectric process an X- or γ-ray gives up all its energy to an electron which usually comes from an inner shell of an atom. The ejected electron is called a photoelectron.

The loss of an electron leaves a vacancy in the inner shell which is filled by an electron dropping in from an outer shell. This process liberates energy which in the case of elements of high atomic number appears as a soft X-ray photon which (unless it escapes from the sample) may in turn undergo photoelectric absorption. In the case of elements of low atomic number the energy is more likely to be used up in ejecting another electron from the same atom. The latter process is known as the Auger effect and the ejected electron is called an Auger electron. The process for low atomic number elements is shown in Fig. 2.1.

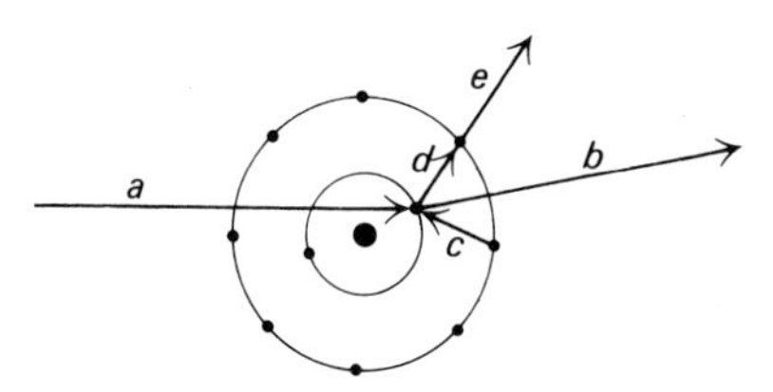

Fig. 2.1 Typical photoelectric absorption process. (*a*) photon enters atom (*b*) electron ejected from inner shell: kinetic energy equals energy of photon minus binding energy (*c*) vacancy in inner shell filled by electron dropping from outer shell (*d*) virtual X-ray collides with electron in outer shell (*e*) electron ejected from outer shell (Auger electron): kinetic energy equals energy of virtual X-ray minus second ionization potential of atom.

If we consider a narrow beam of photons of intensity I photons per cm^2 per second falling on matter, then the change in number of photons per cm^2 per second in a distance dx is given by:

$$dI = -\tau\, I dx \tag{2.1}$$

where τ is the linear attenuation coefficient for the photoelectric process in cm^{-1}. Integrating we obtain:

$$I_x = I_0 \exp(-\tau x) \tag{2.2}$$

where I_x is the intensity at distance x and I_0 the initial intensity.

Equation 2.1 may also be put in the form:

$$-\frac{E}{\rho}\frac{dI}{dx} = \frac{\tau}{\rho} IE \tag{2.3}$$

where E is the energy of each photon and ρ is the density of the medium in g cm^{-3}. The quantity on the left hand side is the energy absorbed from the photon beam per gram of matter per second. The quantity τ/ρ is the mass attenuation coefficient for the photoelectric process in cm^2 g^{-1}.

τ/ρ can be calculated theoretically or estimated from various empirical equations. The variation of τ/ρ with photon energy for two materials is indicated in Figs. 2.2 and 2.3 [4]. τ/ρ falls off approximately as the inverse cube of the photon energy but with higher atomic number elements there are occasional sudden rises in absorption, called 'absorption edges',

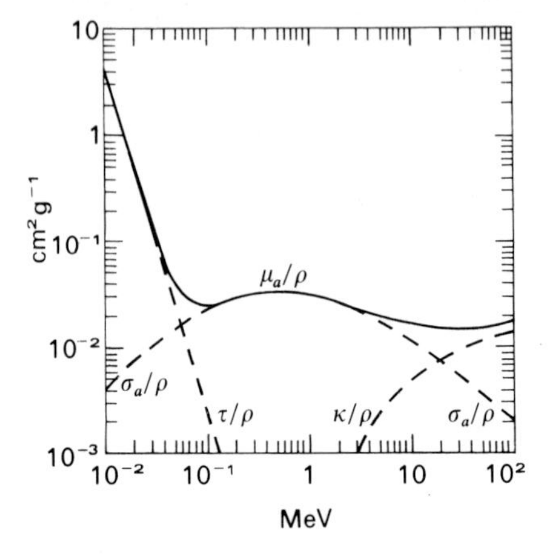

Fig. 2.2 Mass attenuation and absorption coefficients for water as a function of photon energy.

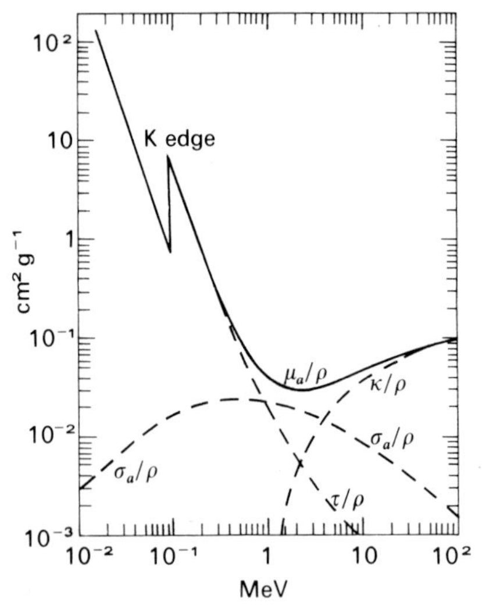

Fig. 2.3 Mass attenuation and absorption coefficients for lead as a function of photon energy.

at the energies where the photons have just enough energy to eject K electrons, L electrons and so on. Above the absorption edge, τ/ρ continues to fall approximately with the inverse cube of the energy. Above the energy of all the absorption edges, τ/ρ increases approximately with the fourth power of atomic number and with the inverse of the atomic weight. Hence photoelectric absorption makes its greatest contribution for low energy X- or γ-rays, and for atoms of high atomic number.

The energy of an X- or γ-ray photon is typically many thousands of electron volts. When a photoelectric event occurs, all of the energy except for the binding energy of the electron is given to the photoelectron. In condensed matter the range of the photoelectrons will be typically less than a millimetre so that, except for those electrons produced at the edge of an irradiated sample, the photoelectrons will normally be rapidly slowed down within the sample, interacting with many more atoms or molecules as they do so. In fact the photoelectrons will usually produce so much change that the original ionizations by the photons themselves will account for a negligible fraction of the change produced. Hence the X- or γ-ray photons will cause most of their chemical or physical effects through the agency of the fast electrons which they generate in the medium.

With very thin samples (for example, thin plastic films), or with gases, most of the energy of the photoelectrons will escape from the sample. However, there will be some electrons which have been generated outside the sample which deposit their energy within it. A sample is said to be in electronic equilibrium with its surroundings if the amount of electron energy generated in the sample but escaping from it is exactly equal to the amount of electron energy generated outside the sample but absorbed in it.

Compton scattering This process was discovered from X-ray scattering experiments performed in the early nineteen twenties. In Compton scattering an X- or γ-ray photon loses part of its energy by ejecting an electron from an atom but, instead of disappearing, is deflected and continues its path with a reduced energy (Fig. 2.4). As in the case of photoelectrons, the ejected electrons (Compton electrons) will usually give energy to the medium, affecting far more atoms or molecules than the original photons as such. The scattered photons will in general escape. The fraction of energy transferred to the medium in the form of kinetic energy of fast electrons is a maximum when the photons are deflected through 180°, under which conditions:

$$\text{energy transferred to medium} = \frac{2E^2}{m_0c^2 + 2E} \qquad (2.4)$$

where E is the initial energy of the photon and m_0c^2 is the rest energy of the electron (0.51 MeV). The average fraction transferred is approximately half the maximum.

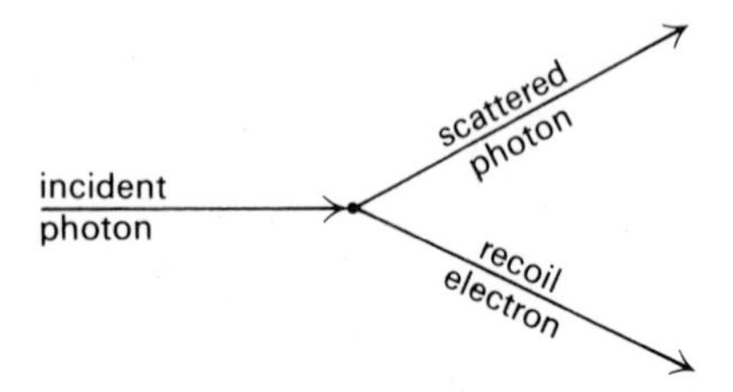

Fig. 2.4 Compton scattering process. In this process the electrons in the medium behave to a close approximation as if they are free. If inner electrons are ejected, further processes take place within the atom, as after photoelectric absorption (see Fig. 2.1).

The number of photons per cm^2 per second which have been scattered in a distance dx is given by an equation similar to that for the photoelectric effect:

$$dI = -\sigma I dx \tag{2.5}$$

where dI is the number of photons per cm^2 per second which have undergone Compton scattering in a distance dx, and σ is the linear Compton attenuation coefficient in cm^{-1}. The mass Compton attenuation coefficient is σ/ρ. Since we are interested in energy *absorbed* in the medium, we may define a mass absorption coefficient, σ_a/ρ, equal to σ/ρ multiplied by the average fraction of energy transferred to Compton electrons. The energy absorbed per gram per second from the photon beam in the form of kinetic energy of Compton electrons is then given by an equation analogous to 2.3:

$$-\frac{E}{\rho}\frac{dI}{dx} = \frac{\sigma_a}{\rho} IE \tag{2.6}$$

σ_a/ρ can be computed using the theoretical formula published in 1929 by Klein and Nishina [5]. σ_a/ρ is a complex function of photon energy and is proportional to the number of electrons per gram of medium. The variation of σ_a/ρ with energy for two materials is included in Figs. 2.2 and 2.3.

Pair production In this process the X- or γ-ray photon, in passing close to the nucleus, becomes converted into a positron-electron pair. This is an example of energy being converted into matter. For the process to be possible it is necessary for the photon energy to be greater than the energy equivalent of the rest masses of the two particles, that is, more than 1.02 MeV. The total kinetic energy of the positron and the electron produced is then equal to the energy of the photon minus 1.02 MeV. The positrons and electrons lose energy in the medium, leading to chemical or physical change, and the positron is ultimately destroyed by combining with an electron, usually to give two photons each of energy 0.51 MeV (annihilation radiation). The pair-production process is shown in Fig. 2.5.

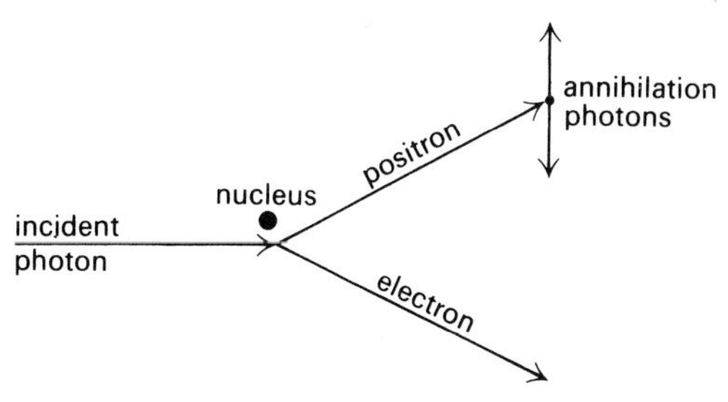

Fig. 2.5 Pair production process.

Like photoelectric absorption and Compton scattering, pair production causes the photon intensity to fall off exponentially:

$$dI = -\kappa I dx \tag{2.7}$$

where κ is the linear attenuation coefficient for pair production. κ/ρ is the mass attenuation coefficient. Another coefficient, κ_a/ρ may also be defined, equal to the attenuation coefficient multiplied by the fraction of energy which is transferred in the form of kinetic energy of electrons and positrons, that is, $(E - 2m_0c^2)/E$. According to theoretical calculations, κ/ρ and κ_a/ρ rise with energy above the 1.02 MeV threshold (see Figs. 2.2 and 2.3) and also with atomic number.

Total absorption coefficient Consider a sample whose dimensions are small compared with the penetration of the incident X- or γ-rays and which is in electronic equilibrium with its surroundings.† The total energy absorption per gram of sample per second due to the interaction by all three processes is given to a close approximation by:

$$-\frac{E}{\rho}\frac{dI}{dx} = \frac{\mu_a}{\rho} IE \tag{2.8}$$

† Where necessary, it is sometimes possible to achieve electronic equilibrium in practice by placing the sample inside a sufficiently thick layer of similar material.

where μ_a/ρ is a mass absorption coefficient given by:

$$\mu_a/\rho = \tau/\rho + \sigma_a/\rho + \kappa/\rho \tag{2.9}$$

κ/ρ is used instead of κ_a/ρ since with large samples of high atomic number (for example 1 cm^3 of lead) most of the annihilation radiation is absorbed in the sample instead of escaping. At very high photon energies κ/ρ and κ_a/ρ are practically equal for all samples, while at photon energies only just above 1.02 MeV the contribution of pair production to the total absorption process is always negligible. With small samples of high atomic number the use of κ/ρ instead of κ_a/ρ may, under certain circumstances, introduce an error as high as about 15 per cent.

Equation 2.8 ignores the production of bremsstrahlung from the fast electrons. This can cause a significant reduction in the amount of energy absorbed by material of high atomic number from beams of very high energy photons but is negligible under most other circumstances.

Table 2.1 *Mass absorption coefficients for photons of various energies* [6]

Photon Energy (MeV)	μ_a/ρ (cm^2 g^{-1}) H	C	N	O	Na	Al	P	S	air	water
0.01	0.00991	1.98	3.38	5.39	14.9	25.6	40.1	50.1	4.63	4.79
0.015	0.0110	0.538	0.908	1.44	4.20	7.48	11.9	15.0	1.27	1.28
0.02	0.0136	0.208	0.362	0.575	1.70	3.06	4.93	6.24	0.512	0.512
0.03	0.0186	0.0596	0.105	0.165	0.475	0.868	1.39	1.77	0.148	0.149
0.04	0.0231	0.0307	0.0494	0.0734	0.199	0.357	0.573	0.729	0.0669	0.0678
0.05	0.0271	0.0234	0.0319	0.0438	0.106	0.184	0.293	0.372	0.0406	0.0419
0.06	0.0305	0.0212	0.0256	0.0322	0.0669	0.111	0.173	0.218	0.0305	0.0320
0.08	0.0362	0.0205	0.0223	0.0249	0.0382	0.0562	0.0820	0.101	0.0243	0.0262
0.10	0.0406	0.0216	0.0224	0.0237	0.0297	0.0386	0.0511	0.0610	0.0234	0.0256
0.15	0.0481	0.0246	0.0248	0.0251	0.0260	0.0286	0.0323	0.0357	0.0250	0.0277
0.2	0.0525	0.0266	0.0267	0.0268	0.0265	0.0276	0.0292	0.0311	0.0268	0.0297
0.3	0.0569	0.0288	0.0287	0.0288	0.0278	0.0283	0.0288	0.0300	0.0288	0.0319
0.4	0.0586	0.0296	0.0295	0.0296	0.0284	0.0287	0.0291	0.0301	0.0295	0.0328
0.5	0.0590	0.0298	0.0297	0.0298	0.0285	0.0288	0.0290	0.0300	0.0297	0.0330
0.6	0.0587	0.0297	0.0296	0.0296	0.0284	0.0286	0.0288	0.0298	0.0296	0.0329
0.8	0.0574	0.0290	0.0289	0.0289	0.0277	0.0279	0.0281	0.0290	0.0289	0.0321
1.0	0.0555	0.0280	0.0280	0.0280	0.0268	0.0270	0.0272	0.0280	0.0280	0.0311
1.5	0.0507	0.0257	0.0257	0.0257	0.0246	0.0248	0.0250	0.0258	0.0257	0.0285
2	0.0465	0.0238	0.0238	0.0239	0.0230	0.0233	0.0235	0.0243	0.0238	0.0264
3	0.0400	0.0210	0.0211	0.0213	0.0208	0.0213	0.0217	0.0225	0.0212	0.0234
4	0.0355	0.0191	0.0194	0.0196	0.0195	0.0201	0.0207	0.0216	0.0194	0.0214
5	0.0320	0.0178	0.0181	0.0185	0.0186	0.0194	0.0201	0.0211	0.0182	0.0200
6	0.0294	0.0169	0.0172	0.0177	0.0181	0.0190	0.0198	0.0209	0.0174	0.0190
8	0.0255	0.0156	0.0161	0.0166	0.0174	0.0186	0.0197	0.0208	0.0162	0.0176
10	0.0229	0.0147	0,0154	0.0160	0.0171	0.0185	0.0197	0.0210	0.0156	0.0168

If we compare two samples exposed to the same photon beam for the same time, the ratio of the energy absorbed per gram for the two samples is in the ratio of μ_a/ρ for the samples. Some values of μ_a/ρ are given in Table 2.1. For samples containing more than one element the overall mass absorption coefficient is given by:

$$\frac{\mu_a}{\rho} = \frac{\mu_{a1}}{\rho_1} w_1 + \frac{\mu_{a2}}{\rho_2} w_2 + \frac{\mu_{a3}}{\rho_3} w_3 + \cdots \tag{2.10}$$

where μ_{a1}/ρ, and so on are the mass absorption coefficients for each element and w_1 and so on are the weight fractions present. For the special case of the irradiation of materials of low atomic number (for example, air, water or hydrocarbons) with hard X- or γ-rays, that is, X- or γ-rays of energy 250 keV to 2 MeV (for example, ^{60}Co γ-rays) the only significant energy absorption process is Compton scattering, so that μ_a/ρ becomes equal to σ_a/ρ and energy absorption per gram in different samples becomes equal to the ratios of the numbers of electrons per gram in the samples.

Other processes Low energy X-rays can interact with matter by coherent scattering but this process is unimportant for radiation chemistry because the energy transferred to molecules is insufficient to cause chemical change. High-energy X-rays (energy generally greater than about 10 MeV) can cause photonuclear reactions, one of the practical consequences of which is to produce some radioactivity, usually of short half-life.

Fast electrons

It has been seen that X- or γ-rays interact with matter to give fast electrons which, on being absorbed in the sample, are responsible for most of any chemical change which is subsequently observed. The same applies to most of the physical effects produced by X- or γ-rays. In fact for most purposes X- or γ-irradiation may be regarded simply as a method of introducing fast electrons into a sample. Fast electrons can also be introduced into a sample directly, for example, from a source of β-particles or an electron accelerator. Fast electrons dissipate most of their energy in matter by causing excitation and ionization.

Excitation and ionization The passage of a fast electron close to an atom or molecule subjects it to an electric impulse which excites or ionizes it. The process may be treated by means of quantum mechanics. An important conclusion of the treatment is that the most abundant activations are those of valence electrons to energies of 10–50 eV [7] (ionization potentials are 10–20 eV). Furthermore the valence electrons of all atoms and molecules are approximately equally liable to be affected. Fast electrons can also produce electronic transitions involving inner shells. Although there are not many such transitions, the energy given up when one happens is very large. However, hardly any of the energy taken up in these transitions is retained in the molecules. Within 10^{-15} s, X-ray photons are emitted or Auger processes occur, the X-ray photons or Auger electrons carrying away a large part of the energy initially imparted to the molecules. The X-ray photons or Auger electrons do not have enough energy to eject electrons from other shells like the ones to which they owed their origin, but they each have enough energy to affect the valence electrons of many more molecules. Consequently the activation of the valence electrons of molecules is even more prominent than already implied. If the activation of outer electrons were to produce little chemical effect, then inner-electron processes could play a significant part in radiation chemistry: such a mechanism has been proposed to account for the production of large numbers of colour centres in the irradiation of the alkali halides (p. 111). But in most cases the processes involving inner electrons are responsible for a negligible proportion of the chemical change.

Absorption of an energy of more than 20 eV or so may often enable more than one molecule to be excited or ionized. If a molecule were to ionize as a result of the absorption of 40 eV for example, the electron emitted would be expected to carry away a significant fraction of the energy. This electron although not sufficiently energetic to escape far from its positive ion, would still have enough energy to excite or ionize another molecule. Effects of this kind are noticed in cloud chambers, where little groups consisting of two or more ionizations can be seen, although single ionizations are more frequent [8]. In condensed phases the groups of events are produced very close together, in clusters or 'spurs' of typical diameter a few nanometres. A very few secondary electrons will have energies of hundreds or thousands of electron volts (for example, the Auger electrons). If their energy is sufficient to enable them to escape far from their site of origin, such electrons may be regarded as distinct particles or 'δ-rays' producing in turn large numbers of excitations or ionizations along their tracks. In between the isolated spurs and the tracks will be blobs where electrons of energy 100–500 eV dissipate all their energy in a limited volume [9]. Fig. 2.6 gives an approximate picture of the distribution of the positive ions produced by absorption of a fast electron in a liquid or solid. The distribution of excitations would be similar.

The transitions induced directly by fast electrons are mainly the ones which are optically allowed. Further details are given in Chapter 4 (p. 59). The electrons with an energy of less than about 100 eV, whether resulting from the deceleration of fast electrons or produced in the medium in spurs, blobs and so on, behave somewhat differently from fast electrons in

100 nm

Fig. 2.6 Part of track of 5 keV electron in condensed matter. Each dot corresponds to one positive ion. Water has 33 million molecules in a 100 nm cube. A solution of concentration 10^{-4}M would contain 60 solute molecules in the cube.

that they can excite to optically forbidden levels. As electrons slow down still further, their energy drops to a level where they can no longer cause excitation of the principal component in the irradiated medium. In this condition the electrons are called 'sub-excitation electrons [10]. Sub-excitation electrons may still be able to excite any minor components. The final process by which electrons lose energy is by increasing the motion of the atoms in the medium. When their energy reaches thermal proportions (about 0.025 eV at room temperature) they may become trapped in the medium or react, for example, with molecules containing electronegative elements or groups, or with positive ions such as the one from which they had come (p. 77).

Stopping power and linear energy transfer The average amount of energy lost by a fast charged particle per unit path length in a medium is known as the stopping power, S. The mass stopping power is S divided by density. For a fast electron of energy E, the stopping power due to excitation and ionization may be expressed by the following equation, originally due to Bethe:

$$S = -\frac{dE}{dx} = \frac{2\pi e^4 N_0 Z}{m_0 v^2}\left[\ln\frac{m_0 v^2 E}{2I^2(1-\beta^2)} - (2\sqrt{1-\beta^2} - 1 + \beta^2)\ln 2 + 1 - \beta^2 + \tfrac{1}{8}(1-\sqrt{1-\beta^2})^2\right] \quad (2.11)$$

where e is the charge on the electron, N_0 is the number of atoms per cm^3 in the medium, Z is the atomic number of the atoms, m_0 is the rest mass of the electron, v is the velocity of the bombarding electron, β is the ratio of v to the speed of light and I is the mean excitation potential of the electrons in the stopping material. β is given by the expression:

$$\beta = \sqrt{1-[m_0c^2/(E+m_0c^2)]^2} \qquad (2.12)$$

I is dependent mainly on the nature and proportion of the atoms in the medium and is comparatively insensitive to the way in which they are combined. Numerous experimental determinations show that for light elements (for example, H, C), the value of I in electron volts is about fifteen times the atomic number, while for heavy elements it is about ten times. Strictly Equation 2.11 applies only to gases; in liquids and solids the energy loss to distant atoms is reduced through polarization of the intervening atoms by the electrical field of the fast electron. The correction for this 'density effect' is most important at high energies, becoming, in the case of water, around 10 per cent at 10 MeV. The stopping power of liquid water as a function of electron energy is shown in Fig. 2.7.

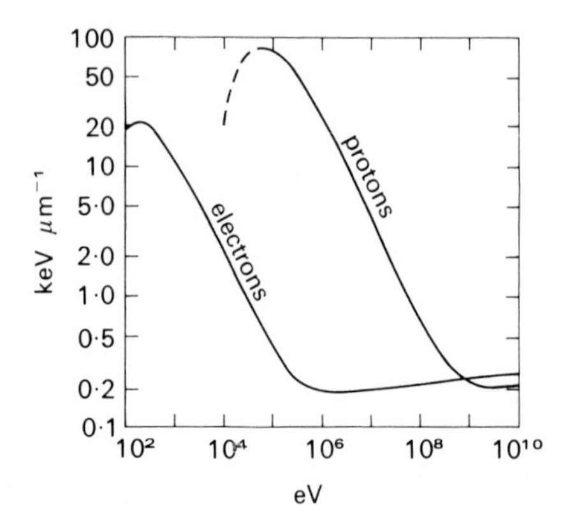

Fig. 2.7 Dependence of stopping power of water on energy of bombarding electrons or protons. While the velocity of the particle is close to that of light, stopping power decreases very slowly with decrease in particle energy. This is the case for electrons at moderate energies, but for protons only at very high energies. Below the relativistic region stopping power continues to increase with decreasing energy down to very low values.

It may be noted that the quantity I appears only in the logarithmic term of the Bethe equation, where its magnitude produces comparatively little effect on *S*. This is consistent with all electrons being roughly equally efficient at taking up energy. It follows that in the irradiation of mixtures the fraction of the total absorbed energy taken up by one component in the mixture is roughly equal to the fraction of the total electrons present which is possessed by that component (its 'electron fraction'). Much of the energy taken up is lost again and taken up by the valence electrons. The fraction of the *numbers* of activated atoms or molecules of the different components in mixtures is to a first approximation equal to the fraction of their *valence* electrons. Another consequence of the Bethe equation is that since the product N_0Z is equal to the number of electrons per cm^3 in the medium, the stopping power *S* is approximately proportional to electron density. The mass stopping power is approximately proportional to the number of electrons per gram.

The value of the stopping power as discussed here is a rather gross mean value for the initial part of the electron track. When the energy of the electron is less than a few hundreds of thousands of electron volts the stopping power will begin to get higher and higher as the electron slows down. Moreover the local value of stopping power in spurs and along δ-ray tracks is very different from the average. In the condensed phases, the manner in which the excitations and ionizations are distributed, as shown qualitatively in Fig. 2.6, is very important for radiation chemistry since, if the excited species or ions are very close together, they, or the species to which they give rise, may react with each other whereas if they are farther apart they will tend to react with the molecules in the medium. A quantity called the 'linear energy transfer' or LET is often used to define the local rate of energy dissipation. LET may be defined as the average energy lost by a fast charged particle due to collisions with energy transfers less than a specified value. If all energy transfers are included the LET may be designated L_∞, in which case it is identical to *S*. To take into account the fact that the energy lost may not be absorbed in the locality (for example if a δ-ray has been ejected) it is more realistic to specify smaller upper limits such as 100 or 500 eV. Use of such quantities enables distributions of energy density along the tracks to be mapped out in some detail.

Other modes of interaction Besides interacting with bound electrons, fast electrons can also interact electrically with the nucleus. When the energy lost from the fast electron appears exclusively as kinetic energy of the recoil atom, the interaction is called an elastic collision. The elastic collisions deflect the electrons (Rutherford scattering) but, because of the large mass of the nucleus, absorb little of their energy. Nevertheless the process can be important in the irradiation of solids. The maximum energy which can be transferred from a fast electron of kinetic energy E is given by:

$$E_m = \frac{2(E + 2m_0c^2)E}{Mc^2} \tag{2.13}$$

where m_0 is the rest mass of the electron, c is the velocity of light and M is the mass of the nucleus. If E_m is greater than a certain threshold value, which from experimental measurements lies within the range 6–80 eV and is typically around 25 eV, then the collision process can displace a struck atom in a solid to a new position. The number of atoms displaced can be calculated [11], and is very small. For instance in the irradiation of germanium, one 3 MeV electron, if completely absorbed, will displace on average only 0.1 atoms. In the irradiation of most chemical systems, the effects produced by excitation and ionization are so large by comparison that displacement effects are not noticed. But in the irradiation of materials like metals and semiconductors, the displacements can give rise to measurable effects.

An electron may also interact with the nucleus to produce bremsstrahlung (p. 8). The process is sometimes called a radiative collision. The main importance of the process is that, as already mentioned for X- and γ-radiation (p. 24), it provides a mechanism by which energy can escape from the medium without producing chemical change.

Fast electrons can also interact with matter to produce low energy electromagnetic radiation called Cerenkov radiation. The process is like the production of the shock wave from supersonic aircraft and it happens whenever the velocity of the electron is greater than the velocity of light in the medium. Cerenkov radiation is responsible for the blue glow seen in swimming pool reactors and round strong sources of γ-radiation stored under water. It does not make a significant contribution to the mechanism of radiation-chemical reactions because very little of the energy of the fast electrons is lost in this way. It is however a consideration in the use of pulse radiolysis equipment (p. 90) with optical means of detection.

Heavy particles

Charged particles Charged heavy particles, for example, protons, deuterons, α-particles or fission fragments, behave like electrons in losing most of their energy in matter in causing excitation and ionization. The excitation and ionization processes are very similar to those with electrons and in fact much of the excitation and ionization is actually caused by electrons produced in the primary ionization processes.

One important difference from fast electrons is that owing to their large mass, heavy particles are travelling much more slowly than fast electrons with the same energy. This increases the value of the stopping power, which is given by an equation which is similar to the non-relativistic form of Equation 2.11:

$$S = -\frac{dE}{dx} = \frac{4\pi e^4 z^2 N_0 Z}{m_0 v^2} \ln \frac{2m_0 v^2}{I} \tag{2.14}$$

where *ze* is the charge on the particle of energy *E*, and *v* its velocity, and the other symbols have the same meaning as in Equation 2.11. The stopping power of liquid water towards protons of various energies is depicted in Fig. 2.7. The linear transfer of energy in condensed matter is often so

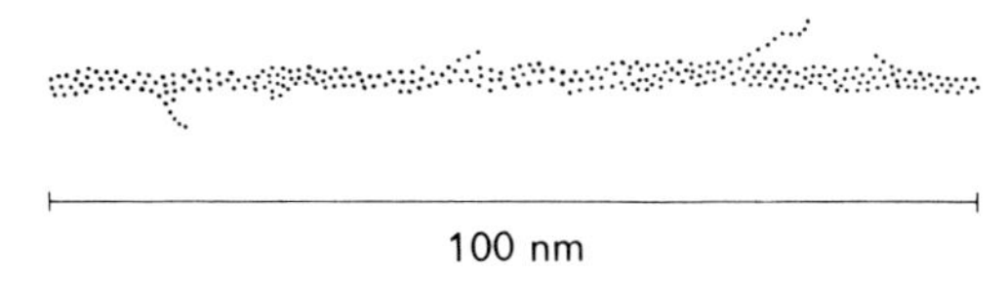

Fig. 2.8 Part of track of polonium α-particle in condensed matter. Each dot corresponds to one positive ion. It is instructive to compare this diagram with Fig. 2.6, noting the difference in scale.

large that the spurs overlap, so that the appearance of a heavy particle track (Fig. 2.8) may be very different from that of a fast electron track (p. 27). Some typical values of LET for charged heavy particles, as well as for fast electrons from various sources, are shown in Table 2.2.

Table 2.2 *Initial stopping power or LET for liquid water*

Particle	*S or L_∞ (keV μm^{-1})*
Theoretical minimum for any charged particle	0.18
2 MeV electrons	0.18
Average Compton electrons from ^{60}Co γ-rays	0.26
Average tritium β-particles	0.65
100 MeV protons	0.65
Typical average electrons from X-rays generated at 220 kV	0.7
Typical average electrons from X-rays generated at 100 kV	1.0
10 MeV deuterons	8.2
30 MeV α-particles	24
Polonium α-particles	90
Uranium fission fragments	1800

Another feature of heavy particle irradiation is that because of the larger mass of the heavy particles, the amount of energy which can be transferred in elastic collisions is much greater than for electrons. The maximum energy which can be transferred to an atom of nuclear mass M is given for a heavy particle of energy E by:

$$E_m = \frac{4MM_1E}{(M+M_1)^2} \tag{2.15}$$

where M_1 is the mass of the particle. The displacement of atoms is important in the irradiation of simple inorganic solids. If the energy transferred is greater than the threshold energy for displacement, a knock-on will be produced and this may be sufficiently energetic to produce more knock-ons, leading to a cascade of displacements. The amount of energy lost in elastic collisions is small compared with that lost in excitation and ionization until the speed of the particle becomes so low that the orbital electrons in the medium are able to adjust themselves to the approach of the particle without being excited or ionized. Beyond this stage elastic collisions become the dominant process. The change-over between the two energy-loss regions occurs when the energy of the particle in eV reaches approximately $M_1/2m_0$. In the irradiation of substances which can be affected by excitation and ionization the effects of elastic collisions are normally proportionately very small but when excitation and ionization are without effect, as is often the case in the irradiation of simple inorganic solids like metals, the elastic collisions become the fundamental processes responsible for most of the radiation effects observed.

When a charged particle has a high energy it can penetrate the electron clouds of atoms, so that the elastic collisions then take place directly with the nucleus without the electrons producing any screening and the Rutherford scattering equations apply to the displacements produced. The mean energy transferred in collisions which displace atoms, $\overline{E}$, is then given by:

$$\overline{E} = E_d \ln \frac{E_m}{E_d} \tag{2.16}$$

where E_d is the threshold energy for displacement (see above). $\overline{E}$ is very much less than E_m. For example, for a 10 MeV deuteron bombarding copper (atomic weight 63.5) $\overline{E}$ would be 270 eV for $E_d = 25$ eV, where from Equation 2.15, E_m is 1.2×10^6 eV. Below an energy E_A, the predominant interaction is with the whole atom ('hard-sphere' collision). E_A is given by:

$$E_A = E_R[2(M+M_1)/M]zZ\sqrt{z^{2/3}+Z^{2/3}} \tag{2.17}$$

where E_R is the Rydberg energy, 13.60 eV, and the other symbols have the meaning already defined. In this region the mean energy transferred per collision is approximately half the maximum, that is, a much greater fraction than in the case of Rutherford collisions. This would be the case in copper for nearly all of the knock-ons produced in bombardment with 10 MeV deuterons.

By making the assumption that the atoms in the medium are randomly distributed, and that each displacement can be regarded as a separate event, the number of displacements produced per unit volume of a solid can be calculated [12]. For crystals [13] the simple theory must be modified to take into account the influence of the lattice structure. In crystals with a close-packed regular structure, the atoms tend to receive their impact from atoms which were initially their nearest neighbours. The struck atoms hand on their energy in turn to the next ones in line, so that movement tends to be focused along particular directions. Moreover once energy has been trapped in a focusing mode, it cannot produce more than one displacement. This diminishes the total number of displacements produced [14]. In open, loosely-packed lattices, moving atoms may be deflected into channels in the crystal and then continue to move along, making glancing collisions. This process, known as 'channelling' [15] profoundly influences the energy loss processes, and also leads to fewer displacements being produced than would be expected on the simple theory.

Atoms which have been displaced by elastic collisions, if they do not return to the vacancies created, ultimately become trapped at interstitial positions. A vacancy-interstitial pair is called a Frenkel defect. Sometimes a moving atom will have just enough energy to displace a lattice atom from its position and, having done so, will fall into the vacancy so created. This process is called a replacement collision and may cause changes in solids which contain more than one kind of atom (see p. 107). In the irradiation of the heavier elements the region in which the cascade of collisions takes place may be very small. An extreme example is seen in the effect of fission fragments on uranium where about 100 MeV may be dissipated within about 400 um. The collisions in such cases cannot be treated as separate events, and the process is more like a local melting. For the case quoted the temperature would rise to about 4000°C. Localized regions of very heavy damage may be called 'displacement spikes [16].' The duration of the high temperature may be about 10^{-11}–10^{-10} s. The final structure in the disordered regions may be quite different from that in the original material, perhaps consisting of different crystalline phases, or amorphous regions, or depleted zones surrounded by sheaths of interstitials.

Neutrons Fast neutrons cannot cause excitation or ionization, but they interact with the nuclei of atoms in elastic collisions. The maximum energy transferred in an elastic collision is given by Equation 2.15. Since neutrons are uncharged, the collisions are of the hard-sphere type. The mean energy transferred in a collision is somewhat less than half the maximum value. It can be seen that energy is most effectively transferred when the atoms in the medium are light, so that in the irradiation of matter containing hydrogen atoms (for example, water or polyethylene) the process responsible for most of the energy dissipation is the production of knock-on protons. These lose most of their energy by causing excitation and ionization as discussed in the last section. When the neutrons are of epithermal energies, a significant fraction of the energy of the knock-on protons is dissipated in causing elastic collisions, so that in this case excitation and ionization are no longer the only important precursors of chemical change.

In the irradiation of inorganic solids, the significant events are the cascades of displacements produced by the primary knock-ons. An estimate of the number of displacements initiated by neutron bombardment can be made by making simple assumptions [12]. More realistic calculations of the number and distribution of displacements can be made by computer techniques, taking into account the structure of the solid and interactions between defects. A result of one such calculation is shown in Fig. 2.9 [17].

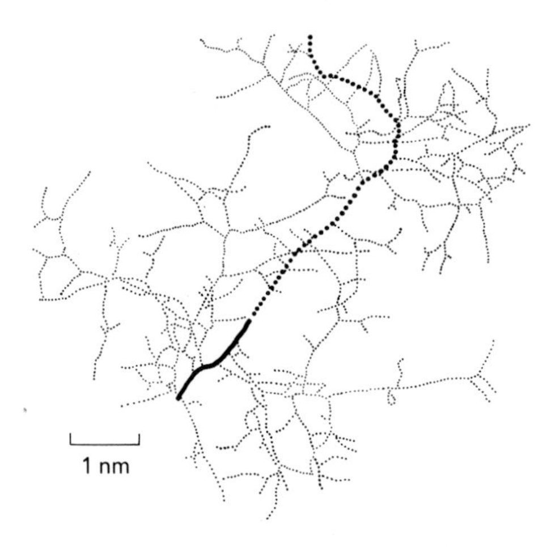

Fig. 2.9 Projection of the tracks of the knock-ons produced in a typical displacement spike in iron. The thick line shows the track of a 5 keV primary knock-on. The dotted lines show the tracks of the secondaries, tertiaries and so on.

Slow neutrons, possessing energies at ordinary temperatures of about 0.025 eV, can interact with matter only by causing nuclear reactions in certain nuclei, for example, ^{14}N (*n, p*) ^{14}C: ^{6}Li (*n*, α) ^{3}H: ^{10}B (*n*, α) ^{7}Li: ^{1}H (*n*, γ) ^{2}H: ^{235}U (*n, f*). The charged particles produced, including the recoil nuclei, cause radiation effects through excitation and ionization and elastic collisions in the ways already discussed.

Nuclear transformations

Many types of radiation are capable of producing nuclear transformations in matter. Besides slow neutrons, energetic-enough X-rays, helium ions and fast neutrons can all cause such effects. In the irradiation of labile chemical substances, the amount of chemical change associated with the nuclear transformation is usually extremely small compared with the amount of change associated with excitation and ionization events occurring in the system. Nuclear transformations are more important in the irradiation of certain inorganic solids, especially where the cross-sections for the processes are high and where the effects of elastic collisions are annealed out by heat (see Chapter 5).

Although the amount of change associated with nuclear transformations is small, it is often possible to study it in detail because of the great sensitivity of nuclear counting equipment. The classical experiment in this field was done by Szilard and Chalmers, who irradiated ethyl iodide with slow neutrons and showed that most of the ^{128}I which was formed by the reaction ^{127}I (*n*, γ) ^{128}I appeared in an inorganic form and could be extracted into an aqueous solution [18]. This could not be because the iodine atoms had been knocked out of the organic molecules by the neutrons, because the C—I bond strength is about 2 eV and the energy of each slow neutron is only 0.025 eV. The reason is that the recoil from the γ-rays provides sufficient energy to break the bonds.

If the energy of a γ-ray is E, then the momentum of the recoiling nucleus, equal to the momentum of the gamma ray, is E/c where c is the velocity of light. The energy of the recoiling nucleus is then $E^2/2Mc^2$ where M is the mass. The mass of the hydrogen atom multiplied by c^2 is equivalent to 931 MeV. The energies of the γ-rays emitted in n, γ reactions are usually several MeV, while bond strengths are usually only a few eV, so that in spite of complications caused by the simultaneous emission of several γ-rays and a tendency to move whole molecules, the recoil energies are often sufficient to break chemical bonds whatever the direction of the recoil. The emission of a single 5 MeV γ-ray for example would lead to a recoil energy of 135 eV for a nucleus of atomic weight 100, while a typical bond strength might be 4 eV.

β-decay also often involves bond breakage, although this can only be detected if the daughter nucleus is radioactive or if the molecule is doubly labelled. β-decay introduces the additional factor that the daughter is a different element from the parent, so that chemical reaction may well be inevitable even in the absence of bond breakage by recoil. In the case of isomeric transitions,† γ-ray energies are only of the order of 100 keV so that bond breakage does not happen. However if an internal conversion electron is emitted the bond is found to break. This is not because of recoil but because the loss of an electron from an inner shell, followed by emission of an Auger electron, leaves the molecule in a highly unstable condition which leads to bond breakage.

The recoiling energetic radioactive atoms, which may well be charged, may react in several different ways, of which the following are among the more important:

(*a*) They may lose their energy in elastic collisions and then recombine with the rest of the molecule, leading to retention of the radioactivity in the target material.

(*b*) They may replace other atoms or groups, even in reactions which are normally endothermic, for example:

$$^{128}I + CH_3I \longrightarrow CH_2{}^{128}II + H \qquad (2.18)$$

Such reactions can be used to synthesize labelled materials.

(*c*) They can enter into exchange reactions, for example:

$$^{128}I^- + I_2 \rightleftharpoons I^- + {}^{128}II \qquad (2.19)$$

(*d*) They can react to give stable molecules, for example, by reacting with oxygen or by dimerization.

Further discussion of the chemical effects of nuclear transformations is outside the scope of this book. For more information the reader should consult publications on radiochemistry.

Comparison of the effects produced by different types of radiation

Where changes occur as a result of excitation and ionization, the most fundamentally significant property of a type of radiation is the value of the linear energy transfer, LET, for the charged particles responsible for the change.‡ Similarly, when elastic collisions are responsible for change,

† An isomeric transition is a process in which a nucleus in a metastable state decays to a state of lower energy with the emission of a γ-ray. Internal conversion is a process in which an unstable nucleus decays by interaction with its external electrons with the emission of an electron.

‡ Strictly the distribution of transfers of energy in the tracks of the particles.

the basic consideration is the distribution of the displacements produced, one extreme being represented by isolated displacements such as may be produced, for example, by bombardment with electrons of suitable energy, and the other by displacement spikes. Where nuclear transformations give rise to change, the nuclear interactions with the material are the important consideration. However, besides these fundamental considerations there are also certain incidental factors which must be borne in mind when comparing the effects produced by different types of radiation. One such factor is rate of energy input: a given amount of radiation energy may produce quite different quantitative effects depending on whether it is delivered quickly or slowly. This is the case in many polymerizations induced by radiation, for example, where the percentage conversion per unit time is often proportional to the square root of the radiation intensity (see pp. 204–206). In such cases a given input of fast-electron radiation, usually delivered at a high rate, would give much less polymer than the same input of γ-ray energy delivered at a lower rate.

Another factor is temperature of irradiation. Certain types of radiation are normally delivered at higher temperatures than others. For example, reactor irradiation or fast-electron irradiation may well cause appreciable heating of the sample and this may lead to differences from, for example, γ-irradiation where the sample may remain at ambient temperature.

Penetration is another factor. 0.5 MeV electrons and 1 MeV γ-rays may produce basically similar effects on a system, but the γ-rays are very much more penetrating, and this may affect the homogeneity of the irradiation. Finally oxygen concentration may be mentioned. Oxygen affects nearly every chemical reaction produced by radiation ('oxygen effect') so that an α-particle or proton irradiation of a polymer film within the vacuum system of a cyclotron, for example, may well give a different response from irradiation with another type of radiation under conditions where oxygen can get into the sample from the air.

PROBLEMS

1. If a water molecule becomes changed chemically when one of its atoms absorbs a 10 keV X-ray photon, how many molecules will be changed by photoelectrons for every molecule changed by the photons as such if the binding energy of the electrons in oxygen is 530 eV and the photoelectrons change water at the rate of four molecules per 100 eV?

[Ans: 379]

2. Estimate the approximate average energy of the Compton electrons produced when cobalt-60 γ-rays interact with matter.

[Ans: 0.52 MeV]

3. A mixture of cyclohexane (50 g) and carbon tetrachloride (50 g) is irradiated with cobalt-60 γ-rays. (*a*) What proportion of the Compton scattering events occurs in cyclohexane molecules? (*b*) What proportion of the energy of the Compton electrons is imparted in the first instance to cyclohexane? (*c*) What proportion of the total number of ionizations and excitations occurs in cyclohexane molecules?

[Ans: (*a*) 54.3 per cent (*b*) 54 per cent (*c*) 67 per cent]

4. Calculate the initial stopping power in liquid water in keV μm^{-1} for (*a*) the Compton electrons of Question 2 (*b*) electrons of energy 10 keV (*c*) protons of energy 100 MeV (*d*) α-particles of energy 10 MeV. Assume I for water is 66 eV.

[Ans: (*a*) 0.26 (*b*) 2.2 (*c*) 0.65 (*d*) 56]

5. (*a*) Use the principles of conservation of energy and momentum to verify Equation 2.15 for the maximum energy E which can be transferred from a particle of energy E and mass M_1 to a stationary atom of mass M. (*b*) Show that Equations 2.13 and 2.15 would give the same result for transfer of energy from a fast electron if it were possible to ignore relativistic effects.

REFERENCES

1. H. A. Bethe and J. Ashkin, in *Experimental Nuclear Physics*, I, ed. E. Segré. Wiley, New York, N.Y., 1953, pp. 166–357, 'Passage of radiations through matter'.
2. U. Fano, in *Radiation Biology*, I (1), ed. A. Hollaender, McGraw-Hill, New York, N.Y., 1954, pp. 1–144, 'Principles of radiological physics'.
3. R. D. Evans. *The Atomic Nucleus*. McGraw-Hill, New York, N.Y., 1955.
4. Based on R. D. Evans, in *Radiation Dosimetry*, 2nd Edn, I, eds F. H. Attix and W. C. Roesch. Academic Press, New York, N.Y., 1968, pp. 93–155, 'X-ray and γ-ray interactions'.
5. O. Klein and Y. Nishina. *Z. Physik*, **52**, 853–68 (1929), 'Über die Streuung von Strahlung durch freie Elektronen nach der neuen relativistischen Quantendynamik von Dirac'.
6. J. H. Hubbell, *Photon Cross Sections, Attenuation Coefficients, and Energy Absorption Coefficients from* 10 heV *to* 100 GeV. NSRDS-NBS 29, Washington, D.C., 1969.
7. R. L. Platzman. *The Vortex*, **28**, 372–85 (1962), 'Superexcited states of molecules, and the primary action of ionizing radiation'.
8. A. Ore and A. Larsen. *Radiat. Res.*, **21**, 331–38 (1964), 'Relative frequencies of ion clusters containing various numbers of ion pairs'.
9. A. Mozumder and J. L. Magee. *Radiat. Res.*, **28**, 203–14 (1966), 'Model of tracks of ionizing radiations for radical reaction mechanisms'.
10. R. L. Platzman. *Radiat. Res.*, **2**, 1–7 (1955), 'Subexcitation electrons'.
11. J. W. Corbett. *Electron Radiation Damage in Semiconductors and Metals*, Academic Press, New York, N.Y., 1966.

12. G. J. Dienes and G. H. Vineyard. *Radiation Effects in Solids*. Interscience, New York, N.Y., 1957, pp. 6–28.
13. R. S. Nelson. *The Observation of Atomic Collisions in Crystalline Solids*. North-Holland, Amsterdam, 1968.
14. R. H. Silsbee. *J. App. Phys.*, **28**, 1246–50 (1957), 'Focusing in collision problems in solids'.
15. O. S. Oen and M. T. Robinson. *App. Phys. Letters*, **2**, 83–5 (1963), 'The effect of channeling on displacement cascade theory'.
16. J. A. Brinkman. *J. App. Phys.*, **25**, 961–70 (1954), 'On the nature of radiation damage in metals'.
17. J. R. Beeler, Jr., in *Radiation Damage in Reactor Materials*. IAEA, Vienna, 1969, **2**, pp. 3–29, 'Computer experiments to predict radiation effects in reactor materials'.
18. L. Szilard and T. A. Chalmers. *Nature*, **134**, 462 (1934), 'Chemical separation of the radioactive element from its bombarded isotope in the Fermi effect'.

Radiation dose and its measurement

The effect of radiation on matter clearly depends on the amount of irradiation. In the early days of radiation science, measurements of irradiation were most often required in connection with the treatment of patients by radiotherapy and it became common to speak of a radiation 'dose'. Measurement of the amount of irradiation is now known as radiation dosimetry. The standard works include the three-volume *Radiation Dosimetry* [1], which is a revision of an earlier handbook, and a series of reports issued by the International Commission on Radiation Units and Measurements [2].

Units

Yield The appropriate measure of irradiation depends on whether the immediate cause of the radiation effect is ionization and excitation, knock-on processes or nuclear transformation. Except in the irradiation of certain inorganic solids, ionization and excitation are almost always the important processes, and the amount of chemical change depends on the total amount of radiation energy deposited in the system, although LET, temperature of irradiation and so on can modify the response to radiation as discussed in the previous chapter. Yields of radiation-induced reactions are expressed as G-values, where G is the number of molecules changed (formed or destroyed) per 100 eV of energy absorbed. In the irradiation of gases it is sometimes less easy to measure the amount of energy absorbed than to measure the number of ions produced in the gas. The yields can then be expressed as ionic yields, M/N, where M is the number of molecules changed and N the number of ion pairs formed in the gas. Before the early nineteen fifties, yields in liquids were often expressed in M/N too, although N was not really known. Such M/N values in liquids should be multiplied by three to give an approximate G-value.

The rad The amount of energy absorbed per unit weight of medium is known as the absorbed dose. The official unit of absorbed dose is the rad, defined in 1953 as 100 ergs per g (that is, 10^{-2} J kg^{-1}). One hundred ergs per gram is equal to 6.242×10^{13} eV per g. The megarad (Mrad) is sometimes used, equal to 6.242×10^{19} eV g^{-1}. Some workers prefer to quote absorbed doses directly in eV g^{-1} or in eV cm^{-3} rather than in rads. At one time a unit called the 'rep' was used. This unit is now obsolete but when found in the older literature may be taken to be about nine-tenths of a rad.

The roentgen When a sample is exposed to a beam of X- or γ-rays the absorbed dose is not only dependent on the radiation, but also on the properties of the medium (see Chapter 2). For some purposes, for example, when quoting the output of an X-ray set or radioactive source, it is useful to have a unit which depends only on the radiation to which the sample is exposed. The unit of exposure is the roentgen, R. By definition, exposure of air to 1 R liberates a total electrical charge of 2.58×10^{-4} coulombs (of one sign) when all the electrons (negative electrons and positrons) liberated by the photons per kilogram of air are completely stopped in air.

$$1\,\mathrm{R} = 2.58 \times 10^{-4}\ \mathrm{C\,kg^{-1}} \tag{3.1}$$

Meaning of the roentgen Consider 1 cm^3 of air at 0°C and 760 torr pressure (weight 0.001293 g) exposed to a short burst of X- or γ-irradiation.† Secondary electrons will be produced. These will have a typical range in air of several centimetres, and will cause ionization and excitation along their path. If all the ions owing their origin to the interaction of the X- or γ-rays with the 1 cm^3 of air could be collected, the exposure would be given by their total positive charge (or negative charge, since these are equal) divided by the weight of the air in 1 cm^3. If the total charge were 3.33×10^{-10}C, the exposure would be 1 R. Now the charge on the electron is 1.60×10^{-19}C, so a charge of 3.33×10^{-10}C corresponds to 2.08×10^{9} ion pairs. It has been found experimentally that the mean amount of X- or γ-radiation energy which is required to produce one ion pair in air

† Note that by definition the roentgen can be used only for photon irradiation.

is equal to 33.7 eV. Consequently when 1 cm^3 of air is given 1 roentgen of X- or γ-radiation, the energy transferred from the photon beam to the medium is $2.08 \times 10^9 \times 33.7$ eV. The absorbed dose is then $2.08 \times 10^9 \times 33.7/0.00129$ eV g^{-1}, that is, 0.87 rad. Of course the energy taken out of the X- or γ-ray beam by 1 cm^3 of air is not absorbed in that 1 cm^3, but fast electrons from absorption of X- or γ-rays in the surrounding air will compensate for this providing the same dose is received by all the air within a distance corresponding to the maximum range of the secondary electrons, that is, providing the sample is in electronic equilibrium with its surroundings.

Although the roentgen is defined in terms of air, the unit can be used to state the exposure of any substance. However, because of the dependence of energy absorption processes on the properties of the medium, different substances exposed to the same number of roentgens will not necessarily absorb the same amount of energy per gram, and therefore will not necessarily receive the same absorbed dose measured in rads. To know the absorbed dose in a given material exposed to 1 roentgen we need to know the ratio of the energy dissipation per gram in the material and in air. This is equal to the ratio of the mass absorption coefficients for the two media (pp. 23–25). For water exposed to hard X- or γ-rays for example, the ratio of the energy absorption per gram for water and air is equal to the ratio of the number of electrons per gram in the two media. Water contains 3.343×10^{23} electrons per gram and air contains 3.007×10^{23} electrons per gram so that when water is exposed to 1 roentgen of hard X- or γ-rays the absorbed dose is $0.87 \times 3.343 \times 10^{23}/3.007 \times 10^{23} = 0.97$ rads. This simple equivalence does not of course hold for water exposed to soft or very hard X- or γ-rays.

Other units As explained in Chapter 2, X- and γ-rays and fast neutrons generally cause their chemical effect through the action of the fast charged particles (electrons or knock-on ions) which they generate in the medium. The energy transferred in the form of kinetic energy of fast charged particles per unit mass is known as the 'kerma'. If a very thin sample were irradiated, the particles escaping could take away a different amount of energy from that deposited by particles entering, so the kerma could be different from the dose in rads. In many other circumstances however the value of the kerma would be the same as the dose in rads.

For the purposes of radiological protection it is useful to have a unit which takes into account the difference in biological effectiveness of different types of radiation. The appropriate unit is called the rem (originally the 'roentgen-equivalent-man'). The rem is equal to the dose in rads multiplied by appropriate modifying factors. The most important consideration is the LET of the radiation. For radiation of LET less than 3.5 keV per μm (such as hard X- or γ-rays) the modifying factor may be taken to be unity. Radiations of high LET can be many times more effective at causing biological damage than low LET radiations and, in the case of fast neutrons, for example, the dose in rems may be as high as 10 or 20 times the dose in rads.

When radiation damage is being caused by knock-on processes, it is necessary to know the number of particles bombarding the sample. If a beam of particles is bombarding a sample perpendicularly, the dose, which may be called the 'fluence', may be expressed by the total number of particles entering the sample per cm^2. The number of particles per cm^2 per second is called the flux or, more correctly, the flux density. If a sample is being bombarded from all directions, as in neutron irradiation in a nuclear reactor, the fluence, in numbers of particles of a specified range of energies per cm^2, is then the number of such particles entering a sphere whose equatorial cross-sectional area is 1 cm^2. Neutron fluences may also be expressed as '*nvt*' where n is the number of neutrons of the specified energy per cm^3, v is their mean velocity, and t is the time of exposure.

In discussing radiation damage to nuclear fuels, the dose may be expressed in terms of the burnup percentage. This quantity may be defined as the percentage of all the atoms present which has been converted to fission products, but other definitions may also be used such as the percentage of fissile atoms which has fissioned. A related concept is fission density, in fissions per cm^3. Another indication of dose is the number of megawatt-days of energy liberated per tonne (1000 kg) of fuel. In the case of a fuel composed of natural or slightly enriched uranium metal, the percentage of uranium atoms which has fissioned is equal to 1.1×10^{-4} times the number of megawatt-days of energy liberated per tonne of the fuel (MWd per tonne). In accurate calculations at high burnup it is necessary to take into account fission in fresh fissile atoms produced by neutron capture in fertile atoms (Pu from U).

Measurement of absorbed dose

The fundamental requirement for this kind of dosimetry is to measure the energy absorbed in the system whose radiation chemistry is being studied. It is not usually possible to measure this directly and so other systems having a known response to radiation are exposed to radiation under similar conditions, and the results are used to calculate the absorbed dose in the system of interest.

Systems used for such purposes are called radiation dosimeters. The most absolute dosimeters measure the heat produced by absorption of radiation energy. Other kinds measure the ionization produced in air. For the chemist one of the most convenient methods is to measure the chemical change in a solution with an accurately known response to radiation dose. In any given situation the method of choice depends on the type of radiation and the nature of the source, the nature of the sample in which one wishes to measure the dose, the magnitude of the dose and the dose-rate, the accuracy required, the experience of the operator and the facilities available. The conceptual and practical difficulties in dosimetry should not be under-estimated. Although the difficulties have been overcome in many cases, there are still situations where it is not at all easy to measure the dose with acceptable accuracy.

Calorimetry One of the features of radiation calorimetry is the low sensitivity, owing to the small amount of heat to be measured. For example a dose of 10^5 rads will raise the temperature of water by only 0.24°C. Because of this, calorimetry has never been in wide use as a practical method of dosimetry, although it has a vital role in absolute dosimetry. The principles of calorimetry are discussed elsewhere [3].

An important instance of the use of calorimetry as an absolute method of dosimetry was its use in 1953 to determine the absorbed dose in the γ-irradiation of a solution of ferrous sulphate in 0.8N sulphuric acid [4]. A diagram of the calorimeter used is shown in Fig. 3.1. After irradiating for a while the concentration of the radiolysis products from the water reached a steady level (Chapter 7), so that any correction due to conversion of radiation energy into chemical change was eliminated. After determination of the rate of energy absorption in the system, the water in the bulb was replaced by a ferrous sulphate solution, and the concentration of ferric ions was determined after irradiation. The dose-rate in rads per minute in the ferrous sulphate solution would be the same as in the water, so the G-value could be obtained. The answer was $G = 15.6$. Although this value differed from numerous other determinations of the yield which were available at the time it was subsequently proved to be correct. The ferrous sulphate system is now in wide use as a chemical dosimeter (see below, p. 50).

Calorimetry can be used to determine the absorbed dose in substances irradiated inside nuclear reactors [5]. Substances in nuclear reactors absorb energy from (*a*) γ-rays, (*b*) fast neutrons and (*c*) radiations produced inside the substance itself by nuclear transformations due to fast, epithermal or slow neutrons. To determine the dose in a substance it is first necessary to compute that part of the dose which is due to nuclear trans-

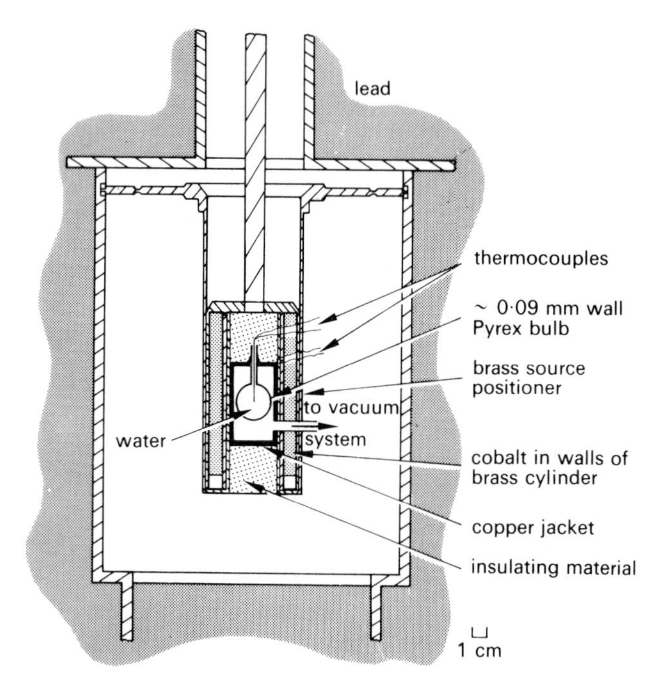

Fig. 3.1 Calorimeter used to measure absorbed dose in cobalt-60 γ-irradiation of water.

formations. This is done from the measured flux of neutrons of the relevant type and the cross section and decay scheme for the reactions occurring, together with the estimated proportion of the energy liberated which would be absorbed in the sample. The calorimeter can then be used to measure the absorbed dose in two materials which absorb energy in one ratio from γ-rays and in a different ratio from fast neutrons. H_2O and D_2O would be suitable materials since they absorb similar amounts of energy from γ-rays but light water absorbs much more energy from fast neutrons than heavy water does (see p. 34). Neither absorbs significant energy from nuclear transformations. The γ-ray energy and the neutron energy absorbed in one of the materials (for example, D_2O) can then be obtained from the measured energy absorption in the two materials together with the ratio of the mass absorption coefficients for the γ-rays present in the reactor and the estimated ratios of neutron energy absorption obtained from the cross sections at the neutron energies present in the reactor, the atomic weights and the mean fraction of kinetic energy transferred per neutron collision. Mean fraction of kinetic energy transferred per neutron collision may be taken to be $2A/(A+1)^2$ where A is the atomic weight of the element (cf. p. 34). The γ-ray energy and the neutron energy in the substance of interest can then be estimated similarly from the energies absorbed in the chosen material. The validity of the method can be tested, and minor corrections incorporated, by carrying out measurements on a variety of materials.

Ionization in gases Small amounts of ionization can be measured very readily so that, in contrast to calorimetry, ionization measurements provide a very sensitive method of measuring dose. Doses of a few millirads can be measured without difficulty. Because of the ease of measurement and their special position with regard to the roentgen, ionization methods have been very extensively used, especially in the radiolysis of gases and for radiological protection and radiotherapy.

One of the simplest examples of ionization methods is in the determination of α-particle doses received from radioactive sources. To do this, one measures with an ionization chamber the total number of ions produced in air by the source in a given time. The energy required to form an ion-pair in air is 35 eV for α-particles, so the energy dissipation in air can be readily calculated. If the source is now used to irradiate any other medium then, provided the whole of the α-particle energy is absorbed in the medium, the rate of energy absorption will be the same as it was in air. In the case of samples whose thickness is less than the range of the α-particles, the stopping power can be used to calculate the energy transferred to the sample.

With more penetrating radiations it is not possible to measure the total ionization produced in air because the range is too great, and so it is necessary instead to measure the ionization produced in a known volume of air. With soft or medium X-radiation (up to a few hundred kilovolts) this method can be used to give an absolute determination of exposure in roentgens. To do this, use is made of a standard ionization chamber (Fig. 3.2). An X-ray beam enters the chamber through a hole and then passes between two parallel plates, one of which is surrounded by guard plates to define the region from which charge is collected. The distance between the hole and the charge collection region must be great enough to stop any secondary electrons arising from interaction of the X-rays with the edge of the hole, and must enable the region to be in electronic equilibrium with the air at either side. The distance between the plates should be greater than the range of the secondary electrons. An electric field of about 50–100 volts per cm is applied between the parallel plates and the change in potential due to the ionization is recorded. If the air is at atmospheric pressure and the temperature is 0°C, then the exposure

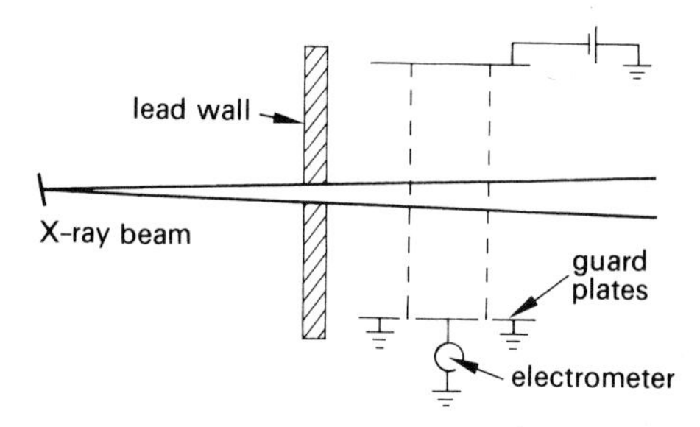

Fig. 3.2 Standard free-air ionization chamber.

at the hole in roentgens is equal to $3.00 \times 10^9 fV/la$, where f is the capacity of the measuring system in farads and V the change in potential in volts, l is the length of the collecting plate in cm, and a is the cross sectional area of the hole in cm^2.

Standard ionization chambers are used for the absolute measurement of dose, for which they are capable of great precision, but they are not suitable for routine use. Moreover they cannot easily be used for hard X- or γ-rays because of the great range of the secondary electrons in air. For routine measurements, cavity ionization chambers are used instead.

A diagram of a typical cavity ionization chamber is shown in Fig. 3.3. In an ideal air-wall chamber the walls would have the same mean atomic number as air and would have the same stopping power for secondary

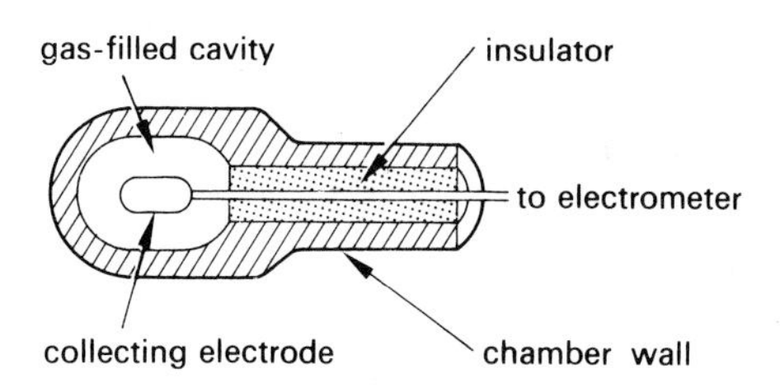

Fig. 3.3 Cavity ionization chamber.

electrons. The walls would be thick compared with the range of secondary electrons. Under these conditions the ionization in the air inside the chamber should be the same as in free air, so that measurement of the ionization in the chamber should give the exposure in roentgens at that point. In practice it is not easy to make a good wall, and the chambers need to be calibrated against radiations of different energy. Providing this is done, air-wall chambers are convenient for measuring exposures in roentgens. Chambers suitable for the accurate measurement of up to

200 or 300 roentgens are commercially available. Chambers are also commercially available for personnel and survey dosimetry. They are less accurate but are robust and convenient in operation.

A variant on the use of the air-wall ionization chamber is the measurement of absorbed dose by means of a chamber with a wall of the same mean atomic number as the medium of interest. If the walls are thick compared with the range of the secondary electrons and the cavity is small compared with the range of the secondary electrons, then the absorbed dose, D_m, in rads or eVg^{-1} may be calculated from the ionization in the gas in the cavity by a relationship based on the principles first discussed by Bragg [6] and Gray [7].

$$D_m = S_{mg} J_g W \tag{3.2}$$

where S_{mg} is the ratio of the mass stopping power of the medium to that of the gas (usually air) for the ionizing particles, J_g is the number of ion-pairs formed per unit mass of gas and W is the mean energy to form an ion-pair in the gas. It is necessary for energy absorption in the walls to be uniform for this relationship to be strictly valid, so that the chamber should not, for example, be too close to a γ-ray source.

The principle of the cavity ionization chamber can be extended in several other ways. One extension is for the measurement of fast neutron doses. For this purpose the hydrogen content of the walls should be close to that of the medium of interest (since fast neutrons interact mainly by producing knock-on protons). The cavity can be filled with air if its dimensions are small compared with the range of knock-on protons in air but, owing to the short range of protons, this limitation is rather severe and it is often better to fill the cavity with a gas with the same composition as the walls. A polyethylene chamber filled with ethylene, for example, can be used to measure the absorbed dose due to fast neutrons in water. Fast neutrons are usually accompanied by X- or γ-rays, and the ionization measurements can be corrected by use of another chamber which is sensitive to X- or γ-rays but less sensitive to neutrons. Another application of the principle is for the measurement of absorbed dose due to fast electron irradiation. For this purpose the walls of the chamber should have a stopping power as similar as possible to that of the material being irradiated. The wall through which the incident electrons enter the chamber should be as thin as possible. The dimensions of the cavity should be small compared with the range of the secondary electrons produced by the incident radiation. Under these conditions the absorbed dose in the medium at the position of the cavity is given by Equation 3.2. Results obtained in this way for water irradiated with fast electrons from a Van de Graaff generator are shown in Fig. 3.4 [8]. The reason for the initial increase in dose below the surface in Fig. 3.4 is that at the surface the medium is being bombarded only by incident electrons, whereas just below the surface it is being bombarded both by incident and secondary electrons. With a very narrow beam of electrons the dose would not increase below the surface because the electrons would be scattered away from the central beam.

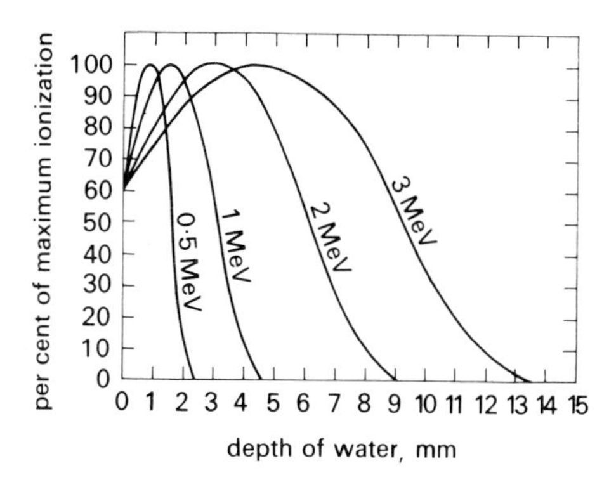

Fig. 3.4 Distribution of dose with depth in irradiation of water with fast electrons of various energies.

Finally the use of ionization measurements for monitoring beams of electrons or positively charged particles produced in machines may be mentioned. The beam is passed through the chamber, which must therefore have thin entrance and exit walls. If the energy of the beam is known, the measurement of the ionization produced, together with knowledge of the stopping power of the gas in the chamber, should give the energy flux in the beam and from this the absorbed dose in the sample could be calculated. Because of scattering, ion recombination and other problems it is not always easy to obtain absolute results using this technique but, if the chamber is calibrated against some other method of dosimetry, the ionization measurements can provide a convenient basis for dosimetry in routine irradiations. A related method of relative dosimetry, suitable for dose-rates where ion recombination makes it impossible to use ionization chambers, is to use an evacuated chamber and measure the emission of secondary electrons from the entrance surface instead of the ionization produced in a gas.

Chemical methods The range of sensitivity of chemical systems to irradiation is such that by measuring the chemical changes produced it is possible to estimate doses from a few rads up to many megarads. Chemical methods of dosimetry are very easy to use compared with other methods, especially for a chemist, and most of them do not require any equipment other than that normally available in a chemical laboratory.

The most reliable and widely used chemical dosimeter is an air-saturated solution of ferrous sulphate in sulphuric acid. The ferrous ions become oxidized to ferric on irradiation, with a G-value which is constant for a given type of radiation whatever the ferrous concentration, dose-rate and irradiation temperature (within wide limits). The system was first proposed for dosimetry in 1927 [9] and is now known as the Fricke dosimeter. The mechanism is discussed in Chapter 7 (p. 156).

To use the Fricke dosimeter it is necessary to prepare a dilute sulphuric acid solution containing approximately 10^{-3} M ferrous sulphate or ferrous ammonium sulphate together with approximately 10^{-3}M sodium chloride. A sulphuric acid concentration of 0.8N (0.4M) was originally adopted by Fricke and Morse so that the absorption of energy from an X-ray beam would be the same function of energy as for air. However, radiation chemists are usually interested in absorbed doses in materials other than air, for example water or hydrocarbons or other materials of low atomic number. When working with X-rays generated at 200–300 kV or less the energy absorption in 0.8N sulphuric acid due to the photoelectric effect is appreciable, and it can be difficult to calculate the absorbed dose in the system of interest from that in the dosimeter unless the energy of the X-rays is known accurately, which is not always the case. The problem can often be avoided by using a 0.1N sulphuric acid solution which for X-rays generated at greater than about 50 kV has an energy absorption which is the same as that of water or at most 2 per cent more. The G-value for oxidation of ferrous ions is about 2 per cent less in 0.1N sulphuric acid than in 0.8N sulphuric acid. When working with cobalt-60 γ-rays, or with other radiations where energy absorption due to the photoelectric effect in sulphur can be neglected, most workers use an acid concentration of 0.8N, merely for traditional reasons.

When preparing the solution it is generally satisfactory to use reagent quality chemicals but the water should be distilled from alkaline permanganate to oxidize the organic impurities which often remain after normal distillation or ion-exchange purification. Organic impurities exert a powerful effect on the performance of the dosimeter and must be rigorously excluded throughout. Solutions are given a dose of about 10 000–20 000 rads, but not more than about 40 000 rads because this would exhaust the oxygen. With inhomogeneous fields care must be taken that the oxygen is not exhausted locally in the solution. The usual method of analysis is the spectrophotometric determination at 304 nm of the ferric ions formed, using unirradiated solution as blank. The molar extinction coefficient of ferric ions in 0.8N sulphuric acid increases by 0.7 per cent per degree between 20° and 30°C. At 25°C it may be taken to be 2205 M^{-1} cm^{-1}. The extinction coefficient in 0.1N sulphuric acid is 1 per cent greater than in 0.8N sulphuric acid. Selected values of the yield of the reaction for several radiations, based on numerous values in the literature, are shown in Table 3.1. Until about 1956 the values of the

Table 3.1 *$G(Fe^{3+})$ in the Fricke dosimeter (in 0.8N H_2SO_4)*

Type of radiation	*$G(Fe^{3+})$*
^{210}Po α-particles	5.2
10 MeV deuterons	9.7
Tritium β-particles	13.0
X-rays generated at 50 kV	13.6
X-rays generated at 100 kV	14.2
X-rays generated at 200 kV	14.6
^{32}P β-particles	15.5
^{137}Cs γ-rays	15.5
^{60}Co γ-rays	15.5
10 MeV electrons	15.5

yield were a matter of controversy, and many workers used to assume the value $G = 20$ for hard X- or γ-rays or fast electrons. Any radiation-chemical yield determined on an incorrect basis can be corrected by dividing by the G-value used and multiplying by the G-value currently thought to be the best.

The absorbed dose received by the Fricke dosimeter in 0.8N sulphuric acid can be calculated from the formula:

$$D_m \text{ (in rads)} = 4.42 \times 10^5[1-0.007(t-20)]OD/G \qquad (3.3)$$

or

$$D_m \text{ (in eVg}^{-1}\text{)} = 2.76 \times 10^{19}[1-0.007(t-20)]OD/G \qquad (3.4)$$

where *OD* is the absorbance (optical density) of the irradiated solution at 304 nm, measured in 1 cm cells with unirradiated solution as blank, *G* is the yield of the reaction for the radiation in use, and *t* is the temperature in °C at which the absorbance was measured. The absorbed dose in the system of interest can be calculated from the dose in the dosimeter.

With suitable modification, the Fricke dosimeter can be used for measurements outside its normal range. For example, the careful use of more sensitive analytical methods permits lower doses to be measured.

If sufficiently large doses are given to the normal Fricke dosimeter in a sufficiently short time (for example, a few thousand rads in a microsecond) the concentration of radicals rises to a level where radical-radical interactions compete with the proper mechanism of the dosimeter so that the G-value decreases: this source of error can be substantially reduced by using 10^{-2}M ferrous instead of 10^{-3}M, leaving chloride out of the solution and saturating with oxygen instead of air.

Numerous other chemical dosimeters are also available. Among those which have received more than purely local acceptance are aqueous solutions containing both ferrous and cupric sulphate, aqueous ceric sulphate, aqueous oxalic acid, cyclohexane, one- and two-phase systems containing chlorinated hydrocarbons, and many others. In pulse radiolysis, transient species formed from water, aqueous potassium thiocyanate or other systems may be measured. In the gas phase, the most widely used dosimeter at present is nitrous oxide, which gives nitrogen and other products on irradiation (see p. 125). When a gas is being irradiated with an external source of X- or γ-rays the absorbed energy usually arises mainly from secondary electrons produced in the walls. Providing the conditions for the validity of the Bragg-Gray relationship are met (p. 48) the energy absorbed per cm^3 in the gas of interest is then equal to the energy absorbed per cm^3 in the dosimeter gas multiplied by the ratio of the stopping powers of gas of interest and dosimeter gas.

Solid state methods The only common feature of dosimeters of this kind is that they are all solids. The parameter measured may consist of ionization, or a change of a 'chemical' type, such as the formation of colour, or a luminescence. Like chemical methods, solid state dosimeters cannot provide an absolute measure of dose but must be calibrated against some other method before they can be used.

The transient ionization produced during the irradiation of semiconductors such as Si or CdS provides one basis for dosimetry. During the irradiation itself the conductivity increases and, after appropriate calibration, conductivity may be used to measure absorbed dose-rate. Owing to their high density and low ionization potential compared to gases, semiconductors are extremely sensitive and the currents which can be measured can be thousands or millions of times those from gases of the same volume. A disadvantage of semiconductors is that their atomic number is very different from that of most of the systems in which one wishes to know the dose, so that with soft or medium X-rays the dose calculation may be subject to large errors. When atoms are displaced in semiconductors (efficiently in heavy particle irradiation, and less efficiently in X-, γ- or electron irradiation) the conductivity changes permanently, and this change occurring in silicon diodes has also been utilized in dosimetry.

Colour changes in plastics can provide an immediately obvious indication of dose. Poly(methyl methacrylate) (Lucite, Perspex, Plexiglas) has received careful study as a quantitative dosimeter [10]. This material becomes yellow on irradiation owing to the development of absorbing species with a maximum near 290 nm. The development of absorption is linear with dose over the region 5×10^4 to about 10^6 rads. The absorption begins to fade within a few hours after irradiation, mainly because of oxygen diffusion into the material, but measurements made soon after the irradiation has finished can provide a convenient measure of dose. Owing to the variability of commercial samples of poly(methyl methacrylate), each batch of material must be calibrated against some other method of dosimetry before use. Changes in the colour of coloured plastics are often used to provide a qualitative indication of dose, for example in checking the precise location of an electron beam or for confirming that foodstuffs being preserved by radiation have in fact received a radiation dose.

Changes associated with colour centres in inorganic solids (p. 109) can also be used for dosimetry. One method uses the radiation-induced thermoluminescence of materials such as aluminium oxide, calcium fluoride, calcium sulphate or lithium fluoride. Lithium fluoride has the advantage of having a mean atomic number which is not too different from that of many of the samples in which one needs to know the dose. The method is sensitive in the range 10^{-2}–10^5 rads. Devices for measuring the light emitted on heating are commercially available. Another method uses the colouration of glass. Ordinary glass is not very suitable because of fading, so special glasses have been developed including silver-activated phosphate glass and cobalt-activated silicate glass. When irradiated silver-activated phosphate glasses are illuminated with ultraviolet light in the 340–360 nm region they emit an orange fluorescence. Radiophotoluminescent dosimeters using this phenomenon have been widely used clinically and in atomic energy installations and have been manufactured for military use by the United States armed services.

Charge collection In the case of irradiation with charged particles from accelerators, the absorbed dose in a target can be calculated if the energy of the particles and the beam current are known. The energy can be determined from the range of the particles or by making use of nuclear reactions with a known threshold energy, or in other ways. The beam current can be measured with the aid of a Faraday cage.

A Faraday cage consists of a target thick enough to stop the beam (the target may consist of the chemical substance being irradiated) which is supported on insulators inside an evacuated chamber (to reduce leakage due to the production of ions by stray radiation). The absorbed dose in the target is then equal to the product of energy of the particles, the beam current and the irradiation time, divided by the target mass. Corrections must be applied for any non-uniformity of particle energy, for back-scattering, for absorption in windows and for energy escaping from the system as bremsstrahlung. With all corrections made the method is capable of great precision. One of the most accurate calibrations of the Fricke dosimeter was made using this method [11]. By making use of the stopping power of a target material, results obtained by charge collection can also be used to calculate the absorbed dose in targets which are too thin to absorb the whole beam.

Counting methods When using α- or β-emitting radioisotopes it is often possible to calculate the absorbed dose in a system if the activity of the source is known. Activity is determined by means of Geiger-Müller counters, scintillation counters or other means. The description of these methods is outside the scope of this book but can be found in numerous other works. The method of dose calculation depends on the way the irradiation is done. It is relatively simple providing the whole of the radiation is absorbed uniformly throughout a system, for example when an α-emitter or relatively weak β-emitter is dissolved in a solution. In this case the energy absorbed per unit time is the product of the number of disintegrations per unit time and the energy per disintegration. The mean energies per disintegration for some common β-emitters, as determined with calorimeters, ionization chambers or theoretical calculations, are given in Table 3.2. If the mean energy of a β-emitter has not been determined it can be assumed as a first approximation to be one third of the maximum energy. Calculations are more complex if the whole of the energy is not absorbed uniformly throughout the system. Such situations are encountered when using radioisotopes for radiotherapy, and the

Table 3.2 *Mean energy of β-particles from some β-emitters*

Isotope	*Maximum β-particle energy (MeV)*	*Mean β-particle energy (MeV)*
^{3}H	0.018 6	0.005 65
^{14}C	0.159	0.050
^{32}P	1.71	0.70

calculations developed for dealing with these situations can sometimes be applied in radiation chemistry.

Measurement of fluence or flux density

Measurement of energy absorption is irrelevant when radiation damage is being caused by knock-on processes, as in the irradiation of metals and other inorganic solids. In such cases the basic requirement is to know the fluence or flux density as well as the energy of the bombarding particles. In irradiation with charged particles, measurement of beam current (p. 53) gives the total number of particles in the beam. As the sample is generally a solid, a separate measurement of the uniformity of the beam across the sample must be made. This can be done by utilizing any physical property which is known to vary in proportion to fluence.

Measurements of fast neutron fluences in reactor irradiations are very difficult indeed and few reactor radiation damage studies have included attempts at precise specifications of the fluences and energies of the neutrons in the reactor. A method which can be used in principle is to measure the number and energy of the recoil protons produced in a hydrogenous medium. A quite different method is to use 'threshold detectors'. These consist of materials whose cross-section is low below a certain threshold energy, above which it remains fairly constant. Some threshold detectors are shown in Table 3.3. Nuclear reactions having a resonance energy can also be used. By use of such reactions it is possible in principle to map out the neutron flux spectrum for the particular

Table 3.3 *Threshold detector reactions* [12]

Reaction	*Energy in MeV at which cross section is about one tenth of its value at high energies*
^{237}Np (*n, f*)	0.6
^{238}U (*n, f*)	1.3
^{58}Ni (*n, p*) $^{58}Co + ^{58}Co^m$	1.9
^{31}P (*n, p*) ^{31}Si	2.0
^{32}S (*n, p*) ^{32}P	2.2
^{27}Al (*n, p*) ^{27}Mg	3.8
^{56}Fe (*n, p*) ^{56}Mn	5.7
^{24}Mg (*n, p*) ^{24}Na	6.4
^{27}Al (*n, α*) ^{24}Na	6.6
^{63}Cu (*n, 2n*) ^{62}Cu	12.0

position in the reactor where the irradiation is being done. For measurement of relative fluences it is possible to utilize properties of materials which change in a known way on irradiation, for example lattice expansion of diamond or conductivity of germanium or silicon.

Thermal neutron fluences are relatively easy to measure. They may be obtained by irradiating a material with a suitable cross-section and half-life and counting the activity produced. Indium, gold and cobalt are frequently used for this purpose.

Burnup percentage in fissile materials is perhaps best measured via radiochemical measurement of selected fission products. For example caesium-137 can be measured. Intact γ-ray spectrometry and isotope dilution techniques followed by mass spectrometer analysis can also be used. Alternatively burnup percentage can be calculated from the heat output measured calorimetrically.

Personnel dosimetry

Neither workers with radiation nor the general public should be exposed to excessive doses of radiation but the question of what is acceptable is a complex matter, involving a balance between the benefits indirectly associated with exposure and the presumed harmful effects of the dose itself. Numerous national and international bodies have recommended limits based on the natural background level of radiation and on radiobiological and other data. The recommendations are always subject to review. As a guide, the dose should be kept to the lowest practicable level and in any case no worker with radiation should receive a whole body dose of more than 5 rem of radiation per year, not more than 3 rem of which may be delivered in any one period of thirteen consecutive weeks. The dose to members of the public and to special categories of workers such as pregnant women, should be kept to substantially lower levels.

It is usual for radiation workers to carry a personal dosimeter. The oldest established type consists of a piece of photographic film wrapped in black paper. An examination of the film with a photometer is used to reveal the dose to which the worker was exposed. Other dosimeters include pocket ionization chambers operating on the cavity principle discussed above (p. 47) and glass, lithium fluoride or other solid state dosimeters.

PROBLEMS

1. Samples of (*a*) aniline ($C_6H_5NH_2$) and (*b*) cyclohexane (C_6H_{12}) are exposed to 10^5 roentgens of X-rays of effective mean energy 100 keV. Calculate the dose in rads for each.

[Ans: (*a*) 86 000 (*b*) 91000]

2. (*a*) Assuming 200 MeV are liberated when nuclear fission takes place, verify the statement on p. 43 that for natural or slightly enriched uranium metal, the percentage of uranium atoms which has fissioned is equal to 1.1×10^{-4} multiplied by MWd liberated per tonne of the fuel. (*b*) For a fuel consisting of UO_2, how many MWd per tonne of fuel will have been liberated if 0.1 per cent of all the atoms present have fissioned? (*c*) How many fissions per cm^3 will this correspond to if the density of the fuel is 10.5 g per cm^3?

[Ans: (*b*) 2.5×10^3 (*c*) 6.9×10^{19}]

3. What would be the rise in temperature of mercury given an instantaneous dose of 10^5 rads? (Specific heat of mercury = 0.139 Jg^{-1}).

[Ans: 7.2°C]

4. A sample of graphite in a hole in a nuclear reactor absorbed 15.5×10^{-4} J g^{-1} s^{-1} from γ-rays and 3.5×10^{-4} J g^{-1} s^{-1} from fast neutrons. Assuming the γ-rays interact only by Compton scattering and the average fast neutron scattering cross sections for aluminium and carbon are in the ratio 2.5 to 2, what would be the total rate of heat absorption in a sample of aluminium irradiated under identical conditions if the sample absorbed 1.75×10^{-4} J g^{-1} s^{-1} from the γ-rays from the reaction ^{27}Al (*n*, γ) ^{28}Al, and 7.3×10^{-4} J g^{-1} s^{-1} from the β- and γ-rays from the decay of ^{28}Al.

[Ans: 24.9×10^{-4} J g^{-1} s^{-1}]

5. (*a*) From the definitions of the units, verify Equations 3.3 and 3.4 for the absorbed dose received by the Fricke dosimeter in 0.8N sulphuric acid (density of solution = 1.024 g cm^{-3}). (*b*) If the absorbance at 304 nm of a Fricke solution irradiated for one hour at a certain position near a source of cobalt-60 γ-rays is 0.50 (1 cm cells, 25°C), what would be the dose in rads in a sample of benzene irradiated in the same place for 24 hours?

[Ans: 3.2×10^5]

6. What would be the dose-rate in rads per minute in a sample of tritiated water of specific activity 1 millicurie per gram?

[Ans: 0.20]

REFERENCES

1. F. H. Attix, W. C. Roesch and E. Tochilin eds. *Radiation Dosimetry*, 2nd Edn, Academic Press, New York, N.Y., 1966–69.
2. *International Commission on Radiation Units and Measurements, Reports 10b*. ICRU, Washington, D.C., 1964.
3. J. S. Laughlin and S. Genna, in *Radiation Dosimetry*, 2nd Edn, II, eds F. H. Attix and W. C. Roesch. Academic Press, New York, N.Y., 1966, pp. 389–441, 'Calorimetry'.
4. C. J. Hochanadel and J. A. Ghormley. *J. Chem. Phys.*, **21**, 880–85 (1953), 'A calorimetric calibration of gamma-ray actinometers'.
5. D. M. Richardson, A. O. Allen and J. W. Boyle. *U.N. Intern. Conf. Peaceful Uses Atomic Energy 1st Geneva*, 1955, **14**, 209–12, 'Dosimetry of reactor radiations by calorimetric measurements'.
6. W. H. Bragg. *Phil. Mag.*, **20**, 385–416 (1910), 'The consequences of the corpuscular hypothesis of the γ and X rays and the range of β rays'.
7. L. H. Gray. *Proc. Roy. Soc. (London)*, **A122**, 647–68 (1929), 'The absorption of penetrating radiation'.
8. J. G. Trump and R. J. Van de Graaff. *J. App. Phys.*, **19**, 599–604 (1948) 'Irradiation of biological materials by high energy roentgen rays and cathode rays'.
9. H. Fricke and S. Morse. *Am. J. Roentgenology and Radium Therapy*, **18**, 430–32 (1927), 'The chemical action of Roentgen rays on dilute ferrosulphate solutions as a measure of dose'.
10. J. W. Boag, G. W. Dolphin and J. Rotblat. *Radiat. Res.*, **9**, 589–610 (1958), 'Radiation dosimetry by transparent plastics'.
11. R. H. Schuler and A. O. Allen. *J. Chem. Phys.*, **24**, 56–9 (1956), 'Yield of the ferrous sulphate radiation dosimeter: an improved cathode-ray determination'.
12. *International Commission on Radiation Units and Measurements, Report 13*. ICRU, Washington, D.C., 1969, 'Newton fluence, neutron spectra and kerma'.

Short-lived intermediates

The excitation or ionization of atoms or molecules is followed by a rapid train of processes involving definable intermediates, about which a great deal is known from fields like molecular physics, photochemistry, mass spectrometry and chemical kinetics, as well as from radiation chemistry. This chapter gives a general discussion of the various types of short-lived chemical intermediate produced through excitation and ionization, including their modes of formation, their properties and their reactivity.

Production of intermediates in the primary activation

Let us consider the activation of an isolated molecule by a fast charged particle. Activation is caused by the rapidly varying electric field produced at the molecule when the particle flies past it. Now it can be shown theoretically that the time-dependent electric field experienced by a molecule in such a situation is the same as would be produced by a bombardment with a stream of rather energetic photons. Over the whole approach and departure of the charged particle, the energy of these 'virtual photons' varies from zero up to a maximum which is just short of hv/p where h is Planck's constant, v is the velocity of the electron and p is the distance of closest approach of the particle to the molecule. For example, in the case of a 0.5 MeV electron (velocity 2.6×10^{10} cm per second) passing 1 nm from a molecule, the energy of the virtual photons varies from zero up to nearly 1200 eV. It can also be shown that the number of the virtual photons possessing a given energy is approximately inversely proportional to that energy. Now the extent to which molecules absorb light from a beam of photons is proportional to the extinction coefficient for those photons or, using the concepts of theoretical physics, to the 'differential oscillator strength' for the frequency of the photons.† Hence the number of molecules, N_S, raised to a given state of excitation, is given approximately by:

$$N_S \propto f_S/E_S \tag{4.1}$$

where f_S is the oscillator strength for the transition and E_S is the energy of the transition [1].

It is an important feature of this treatment that the transitions which occur are those which would occur on absorption of sufficiently energetic light, that is, are optically allowed. Since most molecules are singlets in their ground state, the excited states formed are mostly singlets.‡ The treatment does not however apply to *slow* charged particles such as electrons with energies of less than about 100 eV. Little is known about activations produced by such particles, but they are expected to produce a rather small number of excitations to triplet levels as well as some singlets.

By making use of optical absorption coefficients, electron scattering cross-sections and other data, it is possible to calculate the oscillator strength distribution and hence the variation of N_S with E_S for simple molecules. The oscillator strength distribution for ethanol, which is typical, is shown in Fig. 4.1 [2]. It is a general result of such calculations that oscillator strengths are highest for transitions to energy levels 10–50 eV above the ground state. Molecules which contain double bonds, and especially conjugated double bonds, have larger oscillator strengths for low-energy transitions than do saturated molecules but, even here, by far the greatest part of the total oscillator strength is in the 10–50 eV region. It can thus be seen that most of the primary events in the action

† The differential oscillator strength df/dE in units of eV^{-1}, is equal to $3.483 \times 10^{-5}\,\varepsilon$, where ε is the decadic molar extinction coefficient in $M^{-1}\,cm^{-1}$.

‡ The multiplicity of a state is given by $2S+1$, where S is the number of unpaired electrons in the atom or molecule multiplied by the spin of the electrons, $\frac{1}{2}$. Most molecules in their ground state have their electrons in pairs with their spins in opposite directions so that S is zero and the molecules are in the singlet state. However, a molecule may contain two unpaired electrons, with spins in the same direction. From the exclusion principle, such electrons cannot of course be in the same orbital. From the formula $2S+1$, such a molecule is in the triplet state. Molecular oxygen is exceptional for a molecule in being a triplet in its ground state. Transitions between states of different multiplicity are said to be 'forbidden'. However this rule is not absolute, and there is a small but finite probability of such transitions which becomes quite large where a heavy atom (for example, Br, Hg or Xe) is present in a molecule or in its environment.

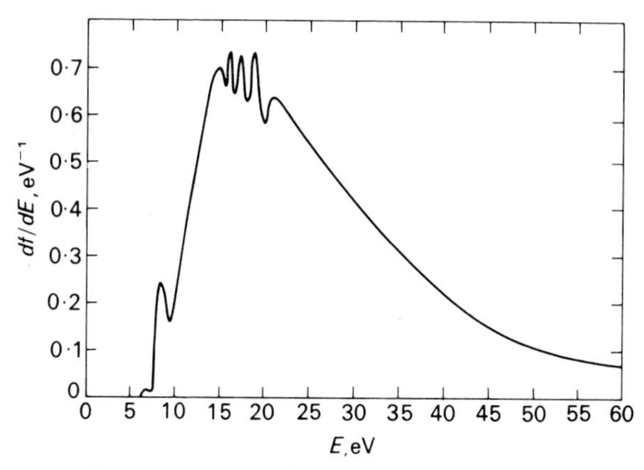

Fig. 4.1 Oscillator strength distribution for ethanol. In an irradiated system the number of molecules activated to an energy *E* is proportional to *df/dE* divided by *E*.

of radiation are to quite highly excited states. In contrast, photochemical excitation is to the lower excited states, for example, a wavelength of 253.7 nm corresponds to only 4.9 eV. The activation process can conveniently be represented on a Jablonski energy level diagram of the type used in photochemistry. Fig. 4.2 shows a simplified diagram for a polyatomic molecule. Activation to the various optically allowed excited states in such diagrams will be produced by fast particles in accordance with the distribution given by Equation 4.1, the oscillator strength being given by curves like that of Fig. 4.1. There will also be a few activations to still higher levels, including activations of inner electrons, but the number of these is very small. There will also be the activations by slow particles to which the theory does not apply. Fig. 4.2 also shows some of the processes occurring after the primary activation (see below).

When a molecule has been excited to a level above its ionization potential (about 10–12 eV for most molecules, see Table 4.1 below) it may ionize in the process. Alternatively it may have a transient existence in a state above the ionization potential, called a superexcited state [3]. In such a state the molecule may lose energy through dissociation as well as by emitting an electron.

As explained in Chapter 2 (p. 26) many of the electrons produced in the ionization process will cause further excitations and, if sufficiently energetic, ionizations. By taking into account all the energetic charged particles in the system from the highly energetic particles forming the initial radiation right down to the sub-excitation electrons, it is possible

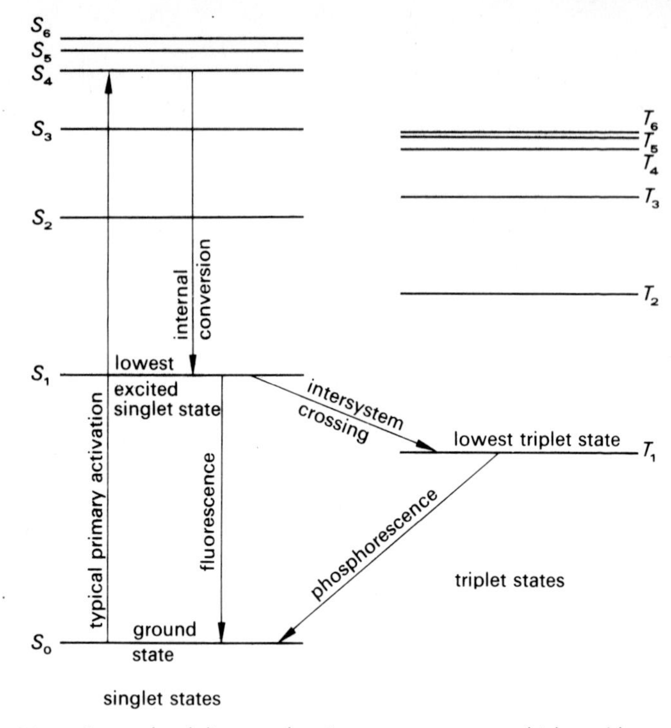

Fig. 4.2 Energy level diagram showing some processes which could occur as a consequence of activation by a fast charged particle.

to estimate from the oscillator strength distribution and other data that in the activation of molecules there will be approximately one excitation produced for every ionization. In the activation of rare gases the ratio will be 1:2.

An important feature of the primary activation of molecules is that it must take place in a time comparable with the electronic oscillation period, that is, 10^{-16}–10^{-15} s. During such a short time the atomic nuclei in the molecule are practically stationary (the time for a molecular vibration is 10^{-14}–10^{-13} s) so that the internuclear distances in the activated state are in the first instance the same as in the ground state. This is known as the Franck-Condon principle, and has been an important feature of photochemistry since 1925–6. It is useful to think of the temporal sequence of events in the action of radiation. If we consider an electron of energy 1 MeV, it will have a velocity of 2.8×10^{10} cm s^{-1} and will therefore traverse a typical molecule of diameter 0.5 nm in 1.8×10^{-18} s. An α-particle of 1 MeV will take 7.3×10^{-17} s. Within 10^{-16}–10^{-15} s of the molecules having been affected by a particle, the primary activation and

Table 4.1 *Ionization potentials* [4]

Species	*Ionization potential (eV)*†	*Species*	*Ionization potential (eV)*
H	13.60	CH_4	12.7
D	13.60	CD_4	12.9
He	24.59	C_2H_2	11.4
Ne	21.56	$C_2H_3\cdot$	8.8
Ar	15.76	C_2H_4	10.5
Kr	14.00	$C_2H_5\cdot$	8.4
Xe	12.13	C_2H_6	11.5
H_2	15.4	C_3H_6	9.7
D_2	15.5	n-$C_3H_7\cdot$	8.1
N_2	15.6	C_3H_8	11.1
O_2	12.1	n-C_4H_{10}	10.6
CO	14.0	c-C_6H_{12}	9.9
HCl	12.7	$CH_2{=}CHCH_2CH_3$	9.6
HBr	11.6	$CH_2{=}CHCH{=}CH_2$	9.1
HI	10.4	$C_6H_5\cdot$	9.2
O_3	12.3	C_6H_6	9.2
CO_2	13.8	$C_6H_5CH_3$	8.8
N_2O	12.9	CH_3OH	10.8
H_2O	12.6	C_2H_5OH	10.5
H_2S	10.4	CH_3COCH_3	9.7
NH_3	10.2	CH_3Br	10.5
CH_2	10.4	CH_3I	9.5
$CH_3\cdot$	9.8		

the Auger events (p. 19) will be complete, and excited states and super-excited states and/or ions will have been produced in the system. Some very approximate times for various processes in radiation chemistry are shown in Table 4.2.

The electronic energy states of solids such as metals and semiconductors differ from those of isolated atoms or molecules in that mutual interaction between the atoms or ions can broaden the electronic states so that they are no longer localized, but rather extend through the solid. The highest levels are conduction bands. In the irradiation of non-conductors, activation to such a level enables the electrons to move through the solid. The 'positive hole' can also migrate, by movement of electrons in the opposite direction. The solid therefore becomes conducting. This activation process is analogous to the ionization of isolated molecules. In certain solids there is a band just below the conduction band, called

† To obtain in kcal per mole multiply by 23.06. To obtain in kJ per mole multiply by 96.49.

Table 4.2 *Approximate time scale in radiation chemistry*

Time (seconds)	*Event*
10^{-18}	1 MeV electron traverses molecule
10^{-16}	1 MeV α-particle traverses molecule
	5 eV electron traverses molecule
	Excitation or ionization by charged particle
10^{-15}	Auger effect
10^{-14}	Autoionization of superexcited states.
10^{-13}	Molecular vibration
	Ion-molecule reaction in liquid or solid
10^{-12}	Time between molecular collisions in liquid or solid
	Internal conversion to lowest excited state in polyatomic molecule
10^{-10}	Ion-molecule reaction in gas at 1 atmosphere
10^{-9}	Time between molecular collisions in gas at 1 atmosphere
10^{-8}	Lifetime for emission of radiation from excited singlet state (fluorescence)
10^{-7}	Lifetime of solvated electron in presence of reactive solute at 10^{-3} M
10^{-5}	Lifetime of solvated electron in presence of reactive solute at 10^{-5} M
10^{-3}	Lifetime for emission of radiation from triplet state (phosphorescence)

the exciton band. Excitation to this level is analogous to the excitation of isolated molecules. An exciton may be regarded as an excited electron bound to a positive hole: it can move through the solid (that is, the electron and the hole travel together) but there is no associated conductivity since the charges do not separate. The primary activations in metals and semiconductors are substantially different from those in the gas phase because of the collective excitations for which there is no counterpart in gases. It is possible that collective excitations may be important in certain condensed molecular media too. If such collective excitations do indeed take place in molecular liquids or solids the 'plasmons' excited are expected to decay to superexcited states within about 10^{-16}–10^{-15} s [5].

Excited species

Even though most of the neutral excited species produced by radiation have much higher levels of electronic excitation than those produced photochemically, the kinds of process by which they dispose of their energy are qualitatively the same for the two cases. The most fundamental source of information about electronically excited species is absorption and emission spectroscopy in the visible and ultraviolet region. Books on spectroscopy or photochemistry should be consulted for fuller information about the chemistry of excited species [6].

Unimolecular processes In the absence of interactions with other species, excited atoms can lose energy only by emitting radiation. Excited *molecules* however, can lose energy of electronic excitation through conversion into vibrational modes, as well as through processes open to atoms. If their energy is greater than the ionization potential they can also lose energy of electronic excitation through ionization, probably in about 10^{-14} s. Little is known about such autoionizations, but it seems likely that the probability may increase with the excess of energy above the ionization potential. The autoionization of excited species is probably a major source of the ions formed in the irradiation of molecules.

In discussing the processes by which neutral excited molecules dispose of their energy, it can be useful to think in terms of the potential energy diagrams which are familiar in photochemistry. A typical diagram for a diatomic molecule is shown in Fig. 4.3. The corresponding diagrams for molecules containing more than two atoms must be represented in more

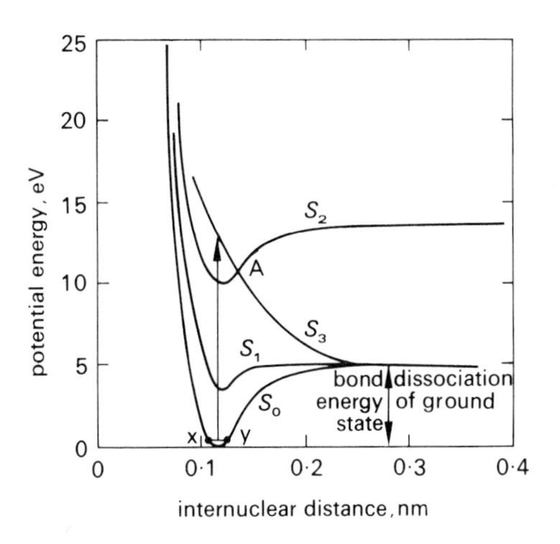

Fig. 4.3 Some typical potential energy curves for a diatomic molecule. Each curve represents the potential energy of the molecule in a particular state of electronic excitation (see also Fig. 4.2). In the ground state the molecule is vibrating between X and Y. The vertical line represents a typical excitation.

than two dimensions and, instead of potential energy curves, we have polydimensional potential energy surfaces. Quantitative values for such curves and surfaces are in general available only for the simpler molecules and for the lower electronic states.

The primary activation takes place in a time which is short compared with the time of a molecular vibration (see above, p. 64). Hence the transition can be represented as a vertical straight line extending from the place on the ground-state curve or surface where the molecule happens to be at the time of the activation. The relative numbers of transitions of the various energies have already been discussed.

Curves of the type shown for S_3 on Fig. 4.3 are dissociative, and any excitation of diatomic molecules to such levels will result in dissociation in the time of a molecular vibration, 10^{-14}–10^{-13} s, unless some other process intervenes (see below). Excitation to states like S_1 or S_2 can result in dissociation if the excitation is to an energy which is higher than the horizontal part of the curve. If this condition is fulfilled, and if no other process intervenes, excitation to such levels will also result in dissociation within 10^{-14}–10^{-13} s. Where more than two atoms are present dissociation will take longer, because the vibrational energy acquired as a result of electronic excitation is distributed among various degrees of freedom, and dissociation cannot occur until an energy equal to the dissociation energy becomes concentrated in one particular bond. This gives more time for other processes to intervene.

The difference between the energy of excitation and the bond dissociation energy may appear as kinetic energy or excitation energy of the free atom or radical fragments produced. Energetic atoms or radicals are referred to as 'hot' and their reactivity will be greater than that of species which are chemically similar but which do not possess excess energy. However in condensed phases it is possible for free atoms or radicals to lose their energy by collision before they have escaped from the molecules which surround them. They will then be in a favourable position to recombine. Recombination within such a 'cage' is known as the Franck-Rabinowitch effect.

At positions where two potential energy curves or surfaces cross (as at A, Fig. 4.3), the potential energy and internuclear distances in a molecule are similar, irrespective of which curve or surface the molecule has been following up to that point. At such positions the molecule can cross over from one curve to another. The process is called 'internal conversion'.† If a molecule has reached curve S_2 at a potential energy which is above A but is insufficient to dissociate the molecule, then at A it can cross over

† This internal conversion is quite different from the nuclear process with the same name defined in the footnote on p. 36.

to curve S_3, which is dissociative. A dissociation which results from a crossing over in this way is called a 'pre-dissociation'.

In the case of molecules containing more than two atoms, there will be many intersections of the potential energy surfaces for different excited states. Hence there will be many possibilities for dropping from higher to lower excited states. Where the lower states are non-dissociative, the excess energy will appear as vibrational energy. In condensed phases this may be rapidly dissipated in collisions. Internal conversion to the lowest excited state is generally considered to be a fast process, of comparable efficiency to dissociation. According to the Franck-Condon principle the arrangement of atoms in the molecule does not alter during the initial activation but the subsequent processes will entail violent alteration, so that permanent changes in the molecule may ensue as a result.

Internal conversion from the lowest excited state to the ground state is a slow process, so that once the lowest excited state of a molecule has been reached, there will be a relatively long time during which other processes can intervene. One such process is the emission of fluorescent radiation. This process takes 10^{-9}–10^{-8} s. Another process of roughly comparable efficiency is 'intersystem crossing' from the lowest excited singlet state to a triplet state.

Bimolecular processes The unimolecular dissipations of energy of excitation take place in competition with processes which are essentially bimolecular. The lifetime of the higher excited states with respect to unimolecular processes is so short, perhaps 10^{-12} s, that it is not easy for bimolecular reactions to compete. Nevertheless molecules in highly excited states may interact with each other in dense regions of the particle tracks [7] and may also react with solutes if present at sufficiently high concentration. If such processes had not happened, the higher states may have gone down to lower states which may have emitted light. Any process which prevents light being emitted is said to have quenched the emission. If a molecule reaches the lowest singlet or triplet level there will be very much more time for the bimolecular processes to take place.

One of the simplest bimolecular processes is the transfer of excitation energy from one species to another:

$$A^* + B \longrightarrow A + B^* \qquad (4.2)$$

Such a transfer can occur only if the energy of the excited state of B is less than that of A. The transfers may take place through the collision of A^* with B. A collisional process in which an excited species loses excitation energy to an atom or molecule in its ground state is sometimes called a 'collision of the second kind'.† A general requirement for this process is that spin should be conserved. Hence if *A* and *B* are both singlets in their ground state, excited singlet *A* will generally produce excited singlet *B*, and triplet *A* will give triplet *B*. Transfer may also occur at a distance, either because A^* emits fluorescent radiation which is then absorbed by *B* (the so-called trivial process) or, for singlets in condensed phases, by way of a radiationless transition involving a quantum-mechanical coupling between A^* and *B*. This process can occur over distances of 5–10 nm. In solids, excitons too may be involved in the energy transfer process.

Excited molecules can also react chemically with other molecules. One type of reaction, which in the case of aromatic molecules can lead to fluorescence quenching, is addition of the excited molecule to a stable molecule. The stable molecule may be the same chemical species but in its ground state: the product of such a reaction is called an 'excimer'. Or it may be a quite different species such as molecular oxygen. Another typical reaction is transfer of an electron to the excited molecule, and another is abstraction of a hydrogen atom by the excited molecule. Information about such reactions can sometimes be gained from photochemical studies. The study of fluorescence emission after irradiation is particularly valuable for providing information about the excited states produced and the processes by which their energy is dissipated. Fluorescence emission itself is put to important practical use in the scintillation counter [8].

The sorting out in particular systems of the details of the reactions of the excited species produced by radiation is a difficult exercise, fraught with pitfalls. One of the simplest possible cases is the irradiation of helium gas containing small amounts of added substances. When pure helium is irradiated with α-particles it is found experimentally that the amount of energy needed to form an ion pair, conventionally called '*W*', is 46 eV, that is, 2.17 ion pairs are formed per 100 eV. However when certain added substances are present it is found that 32.3 per cent more ions are formed than in pure helium [9]. The effect is attributed to the presence of excited helium atoms. In the absence of any added substance the excited helium atoms all lose their energy through collisions with ground state helium atoms but, when suitable added substances are present, collisions of the second kind can lead to ionization, for example:

$$\mathrm{He^* + Ar \longrightarrow He + Ar^+ + e^-} \tag{4.3}$$

The excited helium atoms involved must be metastable ones because

† A 'collision of the first kind' is a collision in which energy of translation is converted into energy of excitation. The primary activation processes discussed at the beginning of this chapter would come under this heading.

otherwise energy would be lost by photon emission before a collision could take place. Helium is known to have metastable states at 20.6 and 19.8 eV and, since the ionization potential of argon is 15.8 eV, Reaction 4.3 is exothermic to the extent of 4.8 or 4.1 eV, according to the metastable state involved, and is therefore permitted from a thermodynamic point of view.

At a given helium pressure, the reaction in which the metastable helium atoms lose their energy by collision with ground state helium atoms without giving ionization is kinetically of the first order. The reaction in which ions are formed (Reaction 4.3) is a bimolecular one (second order) whose rate is proportional to the product of the concentration of the metastable helium atoms and the added substance. Hence the total yield of ions at a concentration C of added substance, $G(C)$, is equal to the yield of ions which are formed without participation of metastable helium atoms, $G(0)$, plus the yield of metastable helium atoms, $G(\mathrm{ex})$, multiplied by the fraction of excited helium atoms which give ions:

$$G(C) = G(0) + G(\mathrm{ex}) \frac{k_{\mathrm{II}} C}{k_{\mathrm{I}} + k_{\mathrm{II}} C} \tag{4.4}$$

where k_{I} and k_{II} are the rate constants for the reactions of the first and second order respectively.

Rearranging Equation 4.4 we obtain:

$$\frac{1}{G(C) - G(0)} = \frac{1}{G(\mathrm{ex})} + \frac{1}{G(\mathrm{ex})} \frac{k_{\mathrm{I}}}{k_{\mathrm{II}} C} \tag{4.5}$$

According to Equation 4.5, we should obtain a straight line if we plot a graph showing the reciprocal of $(G(C) - G(0))$ against the reciprocal of the concentration. The intercept on the $1/(G(C) - G(0))$ axis should give the reciprocal of the yield of excited helium atoms, and from the slope we can obtain the ratio $k_{\mathrm{I}}/k_{\mathrm{II}}$. Some of the results from Reference 9, calculated on the assumption that $G(0) = 2.17$, are plotted in such a way in Fig. 4.4. It can be seen that a straight line plot is indeed obtained and that the value of $1/G(\mathrm{ex})$ is 0.9. Hence $G(\mathrm{ex})$ is 1.1. The ratio $k_{\mathrm{I}}/k_{\mathrm{II}}$ is equal to 1/6 250 (when the concentration of argon is expressed as a mole fraction), which means that collisions with argon molecules are 6 250 times as efficient at giving ionization as collisions with helium molecules are at causing deactivation.

This method of representing data is of general applicability for any system where a first order reaction of a species produced by radiation is in competition with a second order reaction between the species and another substance. The situation was originally encountered in experiments dealing with the decay of fluorescence of iodine vapour (first order) and its quenching by collisional processes (second order) [10]. It is not of course restricted to cases where the reactive species is an excited one. If the value of either k_{I} or k_{II} is known, plots of the 'Stern-Volmer' type can be used to determine the value of the other. The reciprocal of a first-order rate constant is equal to the mean life of the species concerned, so the method can be used to determine the mean life of a species, sometimes loosely called its 'lifetime'.

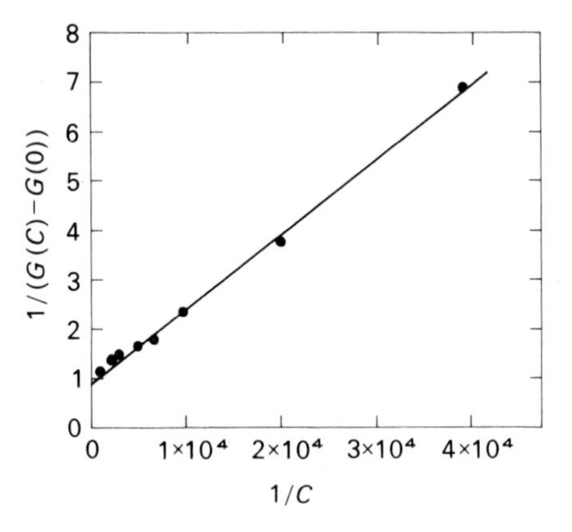

Fig. 4.4 Stern–Volmer plot of yields of ions, $G(C)$, in the α-irradiation of helium containing mol fraction C of argon.

Where an excited species reacts on collision with another species, the maximum possible rate of reaction is equal to the rate at which it encounters its reaction partner by diffusion. This rate constant can be calculated for solutions by making certain simple assumptions [11]: if k is the rate constant for a diffusion-controlled reaction in units of reciprocal moles per litre per second ($M^{-1}\ s^{-1}$) we have:

$$k = \left(2+\frac{r_A}{r_B}+\frac{r_B}{r_A}\right)\frac{2\times 10^3 RT}{3\eta} \tag{4.6}$$

where R is the gas constant in J per mole per degree (8.314), T is the absolute temperature in °K, η is the viscosity of the medium in newton seconds per square metre ($1\ N\ s\ m^{-2} = 1\ J\ s\ m^{-3} = 10$ poise) and r_A and r_B are the radii of the reacting species in any units. This expression is general for diffusion-controlled reactions in solution providing there is no coulombic force between the reactants.

Ions

Positive ions In the first instance, the ions are excited. In the case of simple molecules (for example, H_2) the electronic states of the ions may be known, so that it may be possible to calculate what the unimolecular fragmentation pattern of the ions would be if they were formed in isolation. This is of course the situation encountered in the mass spectrometer where at a low pressure molecules are ionized and then accelerated towards a detector within about 10^{-5} s, so that there is no time for collision with other molecules. In the case of larger molecules no such precision is possible, but for molecules containing more than about six atoms calculations can be made on the basis of the quasi-equilibrium theory [12]. According to this theory ions are formed initially with various amounts of excess electronic and vibrational energy, but do not usually dissociate until at least several vibrations have occurred. During this time, internal conversion will have occurred and the excess energy initially imparted to the ion will be in the form of vibrational energy of the various bonds present. Dissociation or rearrangement of the ion will occur when sufficient energy is present in the appropriate bond or bonds. The fragments may in turn undergo decomposition or rearrangement. A comparison of calculated and experimental fragmentation pattern for propane is given in Table 4.3. An exhaustive compilation of experimental fragmentation

Table 4.3 *Calculated and experimental mass spectra of propane: fraction of total positive ions in the indicated form* [12]

	$C_3H_8^+$	$C_3H_7^+$	$C_3H_5^+$ $C_3H_3^+$ C_3H^+	$C_3H_6^+$ $C_3H_4^+$ $C_3H_2^+$	$C_2H_5^+$	$C_2H_4^+$	$C_2H_3^+$	$C_2H_2^+$	*Others*
Calculated	0.111	0.074	0.056	0.056	0.300	0.185	0.194	0.024	0
Experimental	0.090	0.071	0.103	0.042	0.310	0.183	0.122	0.027	0.052

patterns is published by the American Petroleum Institute [13].

An excited diatomic ion will dissociate in a time close to that required for a molecular vibration, that is, about 10^{-14}–10^{-13} s. Polyatomic ions may take as much as about 10^{-9} s, although experiments with the mass spectrometer show that there are some metastable ions which dissociate while they are being accelerated, and therefore last about 10^{-6} s.

When an ion comes close to a non-polar neutral molecule it will induce a dipole in it. This will produce an attractive force and, if the ion and the molecule come sufficiently close together, capture will occur and the atoms and charge will be in favourable positions for any possible rearrangement to a more stable configuration. The theory for this process was first given by Langevin in 1905. The calculated cross section, σ, for an ion-molecule reaction is given by the expression:

$$\sigma = 2\pi e/v_0(\alpha/\mu)^{1/2} \tag{4.7}$$

where v_0 is the relative velocity of the ion and the molecule, α is the polarizability of the molecule and μ is the reduced mass of molecule and ion. The reaction rate constant in cm^3 per molecule per second is equal to the cross section in cm^2 multiplied by v_0 in cm per second. The equation predicts rate constants as high as about 10^{-9} cm^3 per molecule per second.† Such values are much higher than those for reactions of uncharged molecules because of the attractive forces. Cross sections and hence rate constants can be measured experimentally in mass spectrometers if the ion source is operated at a higher pressure than usual, so that reaction can occur before the ions are accelerated and detected. They can also be measured by a number of other techniques. The values found are often close to the calculated values. Even higher values are found for reactions with polar molecules, where there are additional attractive forces.

Reactions of ions with molecules are of great importance in radiation chemistry. If an ion collides before it has lost the energy associated with the initial excitation it may be at least partly deactivated, or it may undergo a chemical reaction. It is possible that because of the excess energy the reaction may be one which would be forbidden on thermodynamic grounds for reactants of thermal energy. This could be the case in principle in the gas phase at relatively high pressure (in a gas at atmospheric pressure, an ion could react with a neutral molecule in about 10^{-10} s or less) or in the condensed phase. However, thermochemistry provides a generally useful guide to possible reactions. Some data which can be used to work out the gas phase thermochemistry of reactions of ions as well as other species are given in Tables 4.1 and 4.4–4.7. When performing calculations for the liquid phase the heats of solution have also to be taken into account.‡

† 1 cm^3 per molecule per second = 6.02×10^{20} M^{-1} s^{-1}.

‡ The tables are based on experimental determinations and are therefore not entirely internally consistent. Some of the data are based on determinations on isolated molecules, and the thermochemical treatments used to calculate the figures have not been rigorous. Calculations of heats of reaction often involve small differences between large numbers and, in doubtful cases, careful thought

Table 4.4 *Heats of formation (298°K) [4]*

Species	Heat of formation eV	kcal per mole	Species	Heat of formation eV	kcal per mole
H	2.2591	52.095	$C_2H_5\cdot$	1.1	25
D	2.2975	52.981	C_2H_6	−0.8777	−20.24
CO	−1.1455	−26.416	C_3H_6	0.212	4.88
HCl	−0.95672	−22.062	n-$C_3H_7\cdot$	0.958	22.1
HBr	−0.377	−8.70	C_3H_8	−1.076	−24.82
HI	0.275	6.33	n-C_4H_{10}	−1.307	−30.15
O_3	1.48	34.1	c-C_6H_{12}	−1.276	−29.43
CO_2	−4.0785	−94.051	$CH_2{=}CHCH_2CH_3$	−0.001	−0.03
N_2O	0.8504	19.61	$CH_2{=}CHCH{=}CH_2$	1.142	26.33
H_2O	−2.5063	−57.796	$C_6H_5\cdot$	3.1	72
H_2S	−0.214	−4.93	C_6H_6	0.859	19.8
NH_3	−0.4779	−11.02	$C_6H_5CH_3$	0.5182	11.95
CH_2	4.06	93.7	CH_3OH	−2.080	−47.96
$CH_3\cdot$	1.44	33.2	C_2H_5OH	−2.437	−56.19
CH_4	−0.7754	−17.88	CH_3COCH_3	−1.908	−43.99
C_2H_2	2.350	54.19	CH_3Br	−0.36	−8.4
$C_2H_3\cdot$	2.8	65	CH_3I	0.13	3.1
C_2H_4	0.5416	12.49			

One of the simplest reactions between an ion and a molecule is charge transfer, that is, electron transfer. This reaction can take place if the electron affinity of the positive ion (not very different from the ionization potential of the neutral species formed on electron capture) is greater than the ionization potential of the species from which the electron is captured. A simple example is:

$$Ar^+ + CH_4 \longrightarrow Ar + CH_4^+ \qquad (4.8)$$

From the figures in Table 4.1 Reaction 4.8 is exothermic to the extent of 3 eV (70 kcal per mole). If sufficient energy is available (for example, from excitation energy associated with the parent ion) such transfers may be accompanied by dissociation:

$$Ar^+ + CH_4 \longrightarrow Ar + CH_3^+ + H \qquad (4.9)$$

should be given to the reliability of the data used. For further information reference should be made to References 4, 14 or 15 or to other sources of data such as the Handbook of Chemistry and Physics published by the Chemical Rubber Co., Cleveland, Ohio.

Many ions, including parent ions, contain odd numbers of electrons and may be thought of as free radicals. Electron transfer is indeed a typical free-radical reaction (see also below, p. 80). Another typical reaction of such species is hydrogen atom transfer, which can readily occur if the thermochemistry is right, for example:

$$CH_4^+ + C_2H_6 \longrightarrow CH_5^+ + C_2H_5\cdot \tag{4.10}$$

Table 4.5 *Bond dissociation energies* [14]

Bond	*Bond dissociation energy*		*Bond*	*Bond dissociation energy*	
	eV	*kcal per mole*		*eV*	*kcal per mole*
H—H	4.519	104.2	H—iC_3H_7	4.1	94
D—D	4.60	106	H—tC_4H_9	3.9	89
H—H^+	2.69	62.0	H—C_6H_5	4.42	102
H—O	4.436	102.3	H—$CH_2C_6H_5$	3.60	83
H—OH	5.169	119.2	H—CH_2OH	$\leqslant$4.0	$\leqslant$92
H—O_2	2.09	48.2	HO—OH	2.22	51.2
H—O_2H	3.84	88.6	HO—CH_3	3.9	91
H—Cl	4.4718	103.12	C_2H_5—C_2H_5	3.45	79.5
H—Br	3.80	87.6	CH_3—$CH_2C_6H_5$	2.7	63
H—I	3.09	71.3	Cl—Cl	2.51	57.9
H—SH	4.0	92	Br—Br	2.00	46.1
H—NH_2	4.60	106	I—I	1.57	36.1
H—CH_2	3.7	85	Cl—CH_3	3.49	80.5
H—CH_3	4.51	104	Br—CH_3	2.88	66.4
H—C_2H_3	4.55	105	I—CH_3	2.28	52.6
H—C_2H_5	4.26	98.2	I—C_2H_5	2.2	51
H—$CH_2CH{:}CH_2$	3.3	77	I—C_6H_5	2.6	61
H—nC_3H_7	4.1	95			

In this example the product ion CH_5^+ is the Brønsted acid corresponding to CH_4 as base. A Brønsted acid can transfer its proton if the proton affinity of the proton acceptor is greater than that of the neutral species resulting from donation of the proton, for example:

$$CH_4^+ + C_3H_6 \longrightarrow CH_3\cdot + C_3H_7^+ \tag{4.11}$$

From Table 4.6, Reaction 4.11 is exothermic to the extent of 2 eV (50 kcal per mole).

Ions which are hydrogen-deficient, for example, carbonium ions or olefin parent ions, can acquire stability by taking up a hydrogen atom and an electron together, that is, hydride ion transfer, for example:

$$CH_3^+ + C_2H_6 \longrightarrow CH_4 + C_2H_5^+ \quad (4.12)$$

Carbonium ions and olefin parent ions may also form C—C bonds by reaction with neutral molecules, for example:

$$CH_3^+ + CH_4 \longrightarrow C_2H_5^+ + H_2 \quad (4.13)$$

Table 4.6 *Proton affinities* [15]

Species	*Proton affinity*		*Species*	*Proton affinity*	
	eV	*kcal per mole*		*eV*	*kcal per mole*
H	2.6	61	$C_2H_5\cdot$	6.2	143
O	4.5	105	C_2H_6	>5.1	>118
F	3.4	79	C_3H_6	7.4	170
Cl	5.3	122	n-$C_3H_7\cdot$	6.7	155
Br	5.7	131	i-$C_3H_7\cdot$	6.5	149
I	6.3	145	C_3H_8	>5.1	>118
Kr	4.6	106	i-C_4H_8	8.2	189
Xe	4.9	112	t-$C_4H_9\cdot$	6.8	156
H_2	3.0	70	c-C_5H_8	7.5	173
O_2	4.1	94	$C_6H_5\cdot$	8.8	204
OH	6.1	141	C_6H_6	6.5	150
NO	4.7	109	c-C_6H_{10}	7.7	177
HCl	⩾5.2	⩾120	c-$C_6H_{11}\cdot$	7.5	174
HBr	⩾5.6	⩾130	$CH_3O\cdot$	7.1	165
HI	⩾5.1	⩾117	CH_3OH	7.8	180
SH	7.0	161	$C_2H_5O\cdot$	7.5	172
CO_2	⩾5.0	⩾115	C_2H_5OH	8.4	193
H_2O	7.2	167	CH_3OCH_3	8.3	191
H_2S	⩾7.6	⩾175	HCHO	7.0	161
NH_2	7.7	178	CH_3CHO	7.8	180
HO_2	5.4	124	C_2H_5CHO	7.5	172
NH_3	9.1	209	HCOOH	7.0	162
$CH_3\cdot$	5.1	117	CH_3COOH	8.0	184
CH_4	5.2	120	C_2H_5COOH	8.2	190
C_2H_4	6.7	154	CH_3NH_2	>8.8	>202

Several other types of reaction including H_2^- and H_2 transfer have also been observed mass spectrometrically and in other ways. If ions do not react chemically with molecules they may still collect clusters of neutral molecules round them, and such reactions, analogous to solvation in the aqueous phase, have also been observed mass spectrometrically.

The charged products of ion-molecule reactions may react further and in suitable systems chain reactions may occur. The overall pattern of the reactions is governed by competition between unimolecular decomposition or rearrangement and the various possible ion-molecule reactions.

Table 4.7 *Electron affinities* [15]

Species	*Electron affinity*		*Species*	*Electron affinity*	
	eV	*kcal per mole*		*eV*	*kcal per mole*
H	0.8	18	SF_6	1.5	34
O	1.5	34	$CH_3\cdot$	1.1	26
F	3.5	80	C_2H_4	−1.7	−39
Cl	3.7	85	$C_2H_5\cdot$	1.0	23
Br	3.5	80	n-$C_3H_7\cdot$	0.7	16
I	3.1	72	n-$C_4H_9\cdot$	0.7	15
S	2.1	48	$C_6H_5\cdot$	2.2	50
O_2	0.6	13	C_6H_6	−1.5	−35
F_2	2.3	54	$C_6H_5CH_2\cdot$	1.1	25
Cl_2	1.4	32	$(C_6H_5)_2$	−0.5	−12
Br_2	2.1	49	$(C_6H_5)_3C$	2.1	48
I_2	1.8	42	naphthalene	−0.3	−7
OH	1.8	41	anthracene	0.5	12
NO	0.9	21	phenanthrene	0.1	2
CN	2.8	65	pyrene	0.5	12
O_3	3.0	69	$CH_3O\cdot$	0.4	9
N_2O	<1.5	<34	$C_2H_5O\cdot$	0.6	14
NO_2	3.9	91	$CH_3COO\cdot$	3.3	76
NH_2	1.1	26	acetophenone	0.3	8
HO_2	3.0	70	benzoquinone	1.7	39
SO_2	1.0	23	benzaldehyde	0.4	10
CS_2	1.0	23	chloranil	2.6	59
$CCl_3\cdot$	1.4	33	bromanil	2.6	60
CCl_4	2.1	49	iodanil	2.6	59
C_2Cl_6	1.5	34	$CH_3S\cdot$	1.3	31
$SF_5\cdot$	3.6	84	$C_2H_5S\cdot$	1.6	37

Collisional deactivations play an important part. Ultimately the ions will be neutralized. Work with the mass spectrometer and related techniques provides a useful indication of the kind of reactions to be expected in irradiated systems [16] but in the last resort it is necessary to study the system itself, and indeed such studies can be used to give information about ion-molecule reactions which cannot be obtained in other ways [17].

Electrons While the positive ions are decomposing, rearranging or reacting, the associated electrons are slowing down. The slowing down process consists of giving energy to the molecules in the medium, so that in the gas phase, where the density is low, the electrons will travel far from the positive ions. The electrons may react with any substance present which has a positive electron affinity (for example, oxygen). The negative ions formed may react to give further ions. The application of an electric field will allow all the ions to be collected from a gas. If the absorbed dose is known, the energy yield of ions can then be estimated, and expressed either as a *G*-value or, in the inverse sense, as *W*, the amount of energy required to produce an ion-pair. The gas used to measure *W* must be pure, because otherwise extra ions could be formed through reactions like 4.3 (p. 68). Typical values of *W* are given in Table 4.8. If there is no electric field, the electrons or negative ions will end up by neutralizing the positive ions either at the walls or in the gas phase.

In the condensed phase, there are more molecules close to the positive ions than in the gas phase, so that the electrons must lose their energy without travelling so far from the positive ions. Now the distance in centimetres at which the potential energy of a pair of separated ions is equal to the thermal energy kT, where k is Boltzmann's constant in ergs per molecule per degree and T is temperature in degrees K, is given by the Onsager expression:

$$r_c = e^2/\varepsilon_s kT \qquad (4.14)$$

where e is the charge on the electron in esu and ε_s is the static dielectric constant of the medium. The probability of an electron which becomes thermalized at a distance r escaping from its positive ion to become detectable as a free ion is given by $\exp(-r_c/r)$. For liquid water at room temperature where ε_s is 80, the distance r_c is 0.7 nm and most electrons have sufficient energy to enable them to get away from the positive ions. The yield of free ion-pairs in liquid water deduced from chemical experiments is about $G = 3$ (see Chapter 7). For liquid hydrocarbons the dielectric constant is about 2, the distance r_c is therefore about 30 nm, and most electrons normally return to their geminate positive ions probably within about 10^{-10} s. Accordingly the yield of free ion-pairs for hydrocarbons has a low value, generally about 0.1–0.2. However, the yield is sometimes much greater [18] and it is evident that the full story, which is not yet understood, will turn out to contain additional features to those presented here.

When an electron has slowed down in the condensed phase it may become trapped in the medium. If the medium is a solid, the trapped electron may be sufficiently long lived to be studied long after irradiation ('matrix isolation'). One form of trapping, the formation of 'chemical traps', takes place when an electron reacts chemically with a substance possessing a positive electron affinity. Electrons may also be trapped tem-

Table 4.8 *Average energy in eV to produce an ion-pair* [19]

Gas	*W for X- or γ-rays or electrons*	*W for α-particles*
He	41.5	46.0
Ne	36.2	35.7
Ar	26.2	26.3
Kr	24.3	24.0
Xe	21.9	22.8
H_2	36.6	36.2
N_2	34.6	36.39
O_2	31.8	32.3
Air	33.73	34.98
Cl_2	23.6	25.0
Br_2	27.9	—
HCl	24.8	27.0
HBr	24.4	27.0
CO_2	32.9	34.1
N_2O	—	35.7
H_2O	30.1	37.6
NH_3	35	—
CCl_4	25.3	26.3
$CHCl_3$	26.1	—
SF_6	34.9	35.7
CH_4	27.3	29.1
C_2H_2	25.7	27.3
C_2H_4	26.3	28.03
C_2H_6	24.6	26.6
C_3H_8	27.8	—
C_4H_{10}	—	24.8
C_6H_{14}	22.4	—
C_2H_5OH	—	32.6
CH_3Br	28.7	34.6
CH_3I	27.3	—
C_2H_5Cl	25.6	—
C_2H_5Br	25.6	—

porarily at places where the structure of the medium is perturbed. This is the case in the irradiation of such materials as glass and alkali halide crystals at ambient temperatures, and may also be achieved with other solids if the temperature is sufficiently low. Such electrons may be liberated thermally or on illumination with light, and may perhaps move into another trap. In liquids an important kind of trapping is when the electron is solvated. The time for the solvation process is governed by the relaxation time, that is, the time it takes for the solvent molecules to orient themselves round the electric charge. In water at room temperature this may be in the region 10^{-13}–10^{-11} s.

Neutralization The final fate of electrons or negative ions is often neutralization of positive ions. In the gas phase it is not always clear whether this happens homogeneously or at the walls. In the liquid phase, neutralization will be homogeneous, and the rate of neutralization will often be governed by the rate of diffusion of the negatively charged species (negative ions or solvated electrons) and positively charged species (positive ions). The rate constant for the diffusion-controlled reactions of charged species with other charged species is different from that for uncharged species (Equation 4.6) because of coulombic forces, and is conveniently given (in units of $\mathrm{M}^{-1}\,\mathrm{s}^{-1}$) by the expression [11]:

$$k = \frac{4\pi r_{AB} D_{AB} N}{1000} \left\{ \frac{Z_A Z_B e^2}{r_{AB} \varepsilon_s kT} \Big/ \left(\exp\left[\frac{Z_A Z_B e^2}{r_{AB} \varepsilon_s kT} \right] - 1 \right) \right\} \tag{4.15}$$

where r_{AB} is the sum of the radii of the reacting species in cm, D_{AB} is the sum of their diffusion constants in $\mathrm{cm}^2\,\mathrm{s}^{-1}$, N is Avogadro's number, $Z_A e$ and $Z_B e$ are the charges on A and B respectively, where e is the charge on the electron in esu, k is the Boltzmann constant in ergs per molecule per degree and ε_s is the dielectric constant of the medium. If either of the reactants is uncharged the term in braces would become equal to unity and the equation would become:

$$k = \frac{4\pi r_{AB} D_{AB} N}{1000} \tag{4.16}$$

which is closely related to Equation 4.6.

When an electron or negative ion neutralizes a positive ion, energy may be liberated, the amount being somewhat smaller than the ionization potential of the species giving rise to the cation, less, if electron capture has occurred, the electron affinity of the species which has captured the electron. This energy will often give rise to low-lying excited singlet or triplet states of one of the entities involved in the neutralization, and such species will then behave in a similar way to those produced directly by the radiation.

Free radicals

As already implied in this chapter, the excited and ionized species produced by radiation can give rise to free radicals by several different processes. These include:

(*a*) Dissociation of excited molecules into two free radicals, which may or may not initially possess excess energy.
(*b*) Reaction of excited species with other molecules, for example, by electron transfer or hydrogen atom transfer.
(*c*) Unimolecular dissociation of a positive ion-radical produced by the radiation, giving a neutral free radical and an ion possessing only paired electrons.
(*d*) Ion-molecule reactions between ion-radicals and neutral molecules.
(*e*) Capture of an electron by a neutral molecule or by a positive ion possessing only paired electrons.

In addition, solvated or trapped electrons can conveniently be classified as free radicals.

Free radicals vary in reactivity from being so active as to react on collision with almost any stable molecule to being so stabilized by resonance that they can exist indefinitely under ordinary laboratory conditions, as with triphenylmethyl and diphenylpicrylhydrazyl (DPPH). The free radicals produced by radiation may therefore react with suitable stable substances if present in the irradiated medium, or with each other. One common reaction is electron transfer, for example:

$$OH + Fe(CN)_6^{4-} \longrightarrow OH^- + Fe(CN)_6^{3-} \tag{4.17}$$

$$\bar{e}_{aq} + O_2 \longrightarrow O_2^- \tag{4.18}$$

Another typical reaction is hydrogen atom transfer:

$$CH_3 + C_6H_{12} \longrightarrow CH_4 + C_6H_{11}\cdot \tag{4.19}$$

$$Cl + C_6H_{12} \longrightarrow HCl + C_6H_{11}\cdot \tag{4.20}$$

Additions to double bonds or to oxygen are also common:

$$H + C_2H_4 \longrightarrow C_2H_5\cdot \tag{4.21}$$

$$CH_3 + O_2 \longrightarrow CH_3O_2\cdot \tag{4.22}$$

The product of a free-radical reaction is often another free radical, and these usually react further. Sometimes a whole series of reactions is initiated as, for example, in free-radical polymerization. The ultimate fate of free radicals is often disproportionation, for example:

$$2C_2H_5\cdot \longrightarrow C_2H_4 + C_2H_6 \quad (4.23)$$

or dimerization:

$$2CH_3\cdot \longrightarrow C_2H_6 \quad (4.24)$$

$$2I \longrightarrow I_2 \quad (4.25)$$

Thermodynamics is a useful guide to free-radical reactions, although in special circumstances the radicals may have sufficient excess energy to enable them to undergo reactions which would not normally take place because of a high activation energy.

The maximum rates for free-radical reactions can be calculated for the gas phase from simple kinetic theory and for the liquid phase using Equations 4.6, 4.15 or 4.16. Experimental values have been obtained by pulse radiolysis (see p. 90) and other methods, and are found to cover the whole range from the diffusion-controlled rate to zero.

It is interesting to enquire into the number of free radicals produced by a given amount of radiation. In principle this number can be measured experimentally by introducing into the system a substance such as oxygen, iodine or nitric oxide which is known to react with radicals, that is, a radical scavenger. Iodine for example scavenges radicals according to:

$$R\cdot + I_2 \longrightarrow RI + I \quad (4.26)$$

followed by Reaction 4.25. In the gas phase at normal dose-rates, each radical produced by radiation has a much greater chance of meeting a scavenger molecule than another radical, so that determination of the amount of scavenger changed measures the number of radicals produced. However the scavenger may react with excited molecules, ions or electrons as well as radicals, so the interpretation of results is not always unambiguous. In the condensed phases an additional problem arises, since many of the free radicals are formed very close to each other, in spurs or, in the case of high LET radiation, in the tracks of the particles (Fig. 2.6, p. 27 and Fig. 2.8, p. 31). These radicals will react very rapidly with each other, and cannot easily be intercepted except with impracticably high concentrations of scavengers. However, many of the radicals will escape from the spur or track and, once they do so, will be much more likely to react with any solute molecule present than with other radicals. In the irradiation of dilute aqueous solutions with γ-rays for example, there is always a certain fixed yield of hydrogen peroxide and hydrogen gas which must have originated at least in part in radical combination in the spurs, but it is also very often found that the number of solute molecules changed for a given radiation dose is approximately constant over a wide range of solute concentrations (say, 10^{-5}–10^{-3} M) and this can be understood if there is a certain yield of free radicals which can escape from the spur to react with the solute. This yield, often loosely called the radical yield, can be determined by measuring the change in the solute. Typical yields of such free radicals for low LET irradiation of various liquids are shown in Table 4.9.

Table 4.9 *Yields of radicals in the low LET radiolysis of some liquids* [20]

Compound	*Number of radicals formed per 100 eV*
n-Hexane	4.8–5.9
Cyclopentane	4.6–5.5
Cyclohexane	4.8–5.5
Ethylene	2.6
trans-2-Butene	4.0
Cyclohexene	5.2
Benzene	0.7–1.0
Water	6.25
Methanol	6
Ethanol	5–8
Acetone	4
Ethyl ether	9–12
Acetic acid	4
Methyl acetate	7
Carbon tetrachloride	3–7
Ethyl iodide	4–6
Chloroform	5

A mathematical treatment can be attempted for the formation of radicals and other species in spurs or tracks, followed by their diffusion away from their point of origin. As the species diffuse away they may react with each other, with the products of the interaction and with anything dissolved in the medium. One of the most simple situations would be if only one type of radical, $R\cdot$, were formed by the radiation, and if this could diffuse away from its point of origin and react either with another radical to give a stable product R_2, or with a solute to give a stable product RS. For such a case the rate of decrease of concentration of $R\cdot$ at a distance r from the point of origin would be given by:

$$\frac{\partial C_{R\cdot}}{\partial t} = D_{R\cdot}\nabla^2 C_{R\cdot} - k_{RR}C_{R\cdot}^2 - k_{RS}C_{R\cdot}C_S \qquad (4.27)$$

where $C_{R\cdot}$ is the concentration of $R\cdot$ at distance r at time t,† $D_{R\cdot}$ the diffusion coefficient of $R\cdot$, k_{RR} and k_{RS} the rate constants for dimerization of $R\cdot$ and reaction of $R\cdot$ with S respectively, and ∇^2 is the three dimensional Laplacian operator:

$$\nabla^2 = \frac{\partial^2}{\partial r^2} + \frac{\alpha}{r}\frac{\partial}{\partial r} \tag{4.28}$$

In the case of low LET radiation the spurs are assumed to be spherical: r is the distance from the centre of the spur and α is 2. For high LET radiation the radicals are taken to be initially distributed through a cylinder: r is the distance from the centre of the cylinder and α is 1.

Similarly, the decrease in concentration of S would be given by:

$$\frac{\partial C_S}{\partial t} = D_S\, \nabla^2\, C_S - k_{RS} C_{R\cdot} C_S \tag{4.29}$$

By making certain assumptions about the initial distribution of the radicals it is possible in principle to solve this and similar equations. Most of the work has been with water and aqueous solutions although organic liquids have been treated too. The aim of the work is to produce a theory which is consistent with all the valid experimental data on the nature of the reactive species, the rate constants, diffusion constants, yields, effect of LET, dose-rate, solute concentration and so on. The theory has been developed and refined over the years as more and more sophisticated mathematical techniques, using computers, have been applied and as more and better experimental results have accumulated [21, 22].

Direct study of intermediates

The intermediates encountered in radiation chemistry usually have, almost by definition, lifetimes which are too short to enable them to be studied by the established methods of chemistry. Accordingly much of the information about them is derived indirectly, for example from theory, or by inference from the analysis of reaction products. However, the intermediates can be studied directly by various physical methods.

One way to make the intermediates amenable to study is to increase their lifetime. This can often be done by irradiating in the solid state (if necessary at very low temperatures) so that the intermediates are prevented from reacting. If the irradiation is done in this way the intermediates can be studied after the irradiation has stopped, using optical

† $C_{R\cdot}$ is more strictly defined as the probability density of finding $R\cdot$ at distance r at time t.

absorption spectroscopy, for example, or electron spin resonance. The use of a solid matrix to trap intermediates is sometimes called the 'matrix isolation technique'.

Electron spin resonance This technique can readily be used with irradiated solids, and has also been used for liquids. The principles are discussed here briefly in view of the important part the method has played in radiation chemistry. Textbooks on electron spin resonance should be consulted for further details.

When a free electron is situated in an applied magnetic field, H, there are two possible orientations for its magnetic moment vector. In the orientation of lower energy, the magnetic moment of the electron is parallel with the direction of the applied field, and in the orientation of higher energy it is anti-parallel. The difference between the two energies is equal to $2\beta H$ where β is the fundamental atomic unit of magnetic moment, the Bohr magneton. Transition from the lower to the upper state can be brought about by the absorption of radiation of frequency, ν, where:

$$h\nu = 2.0023\,\beta H \tag{4.30}$$

(the figure 2.0023 arises out of a relativity correction). Pairs of electrons with their spins in opposite directions have no net magnetic moment and so cannot exhibit such absorption phenomena. Accordingly the property of absorbing radiation of appropriate frequency while in a magnetic field is characteristic of substances containing unpaired electrons, for example, transition metals, molecules in triplet states and free radicals.

Electron spin resonance has mainly been used in radiation chemistry to determine the structure of free radicals. The possibility of doing this arises from the fact that if an electron is bound to a magnetic nucleus, or associated with several magnetic nuclei, the fields associated with the nuclei will modify the value of the magnetic field seen by the electron. Consider, for example, an electron bound to a single proton. The magnetic moment vector of the proton, like that of the electron, has two possible orientations in a magnetic field. For one half of the protons the magnetic moment vector will add to the value of H, and for one half the magnetic moment vector will subtract from H, by the same amount. Equation 4.30 may be re-written as:

$$h\nu = g\beta H \tag{4.31}$$

where g is known as the spectroscopic splitting factor. For an electron bound to a single proton, if H is fixed, there will be two values of $h\nu$ where absorption will occur, and these will be symmetrically placed about the value given by g $\sim$ 2. If $h\nu$ is fixed there will be two values of H where absorption will occur. When an electron is associated with several magnetic nuclei a more complicated hyperfine structure will arise. Measurement and analysis of the hyperfine structure gives the information about the structure of the free radical

It can be shown that the sensitivity of an ESR spectrometer rises greatly with frequency and hence with magnetic field strength. In practice use is generally made of microwaves in the Q-band, which centres at 35 000 Mcycles per second and requires 12 500 gauss for $g = 2$, or in the X-band, centring at 9 400 Mcycles per second, with 3 350 gauss for $g = 2$. It would be difficult for practical reasons to operate at a fixed field strength and vary the frequency, so spectrometers operate instead at a fixed frequency, with the field strength being varied over the range of interest. A diagram of simple apparatus is shown in Fig. 4.5. An oscilloscope or

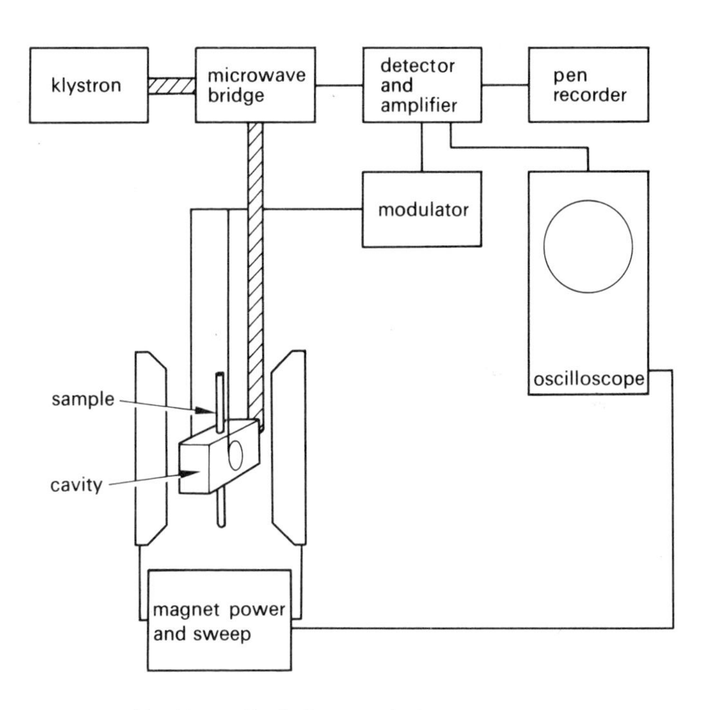

Fig. 4.5 Block diagram of ESR spectrometer.

chart recorder could be used to display ESR absorptions as a function of magnetic field. However, the influence of frequency variations can be eliminated by modulating the magnetic field, in which case the first derivative of the absorption line is obtained. Second derivative curves can also be obtained: these discriminate against broad lines, and also improve the resolution of overlapping hyperfine lines. The relation between absorption, first derivative and second derivative spectra can be seen from Fig. 4.6. Typical second derivative spectra are shown in Figs. 4.7 and 4.8 [23].

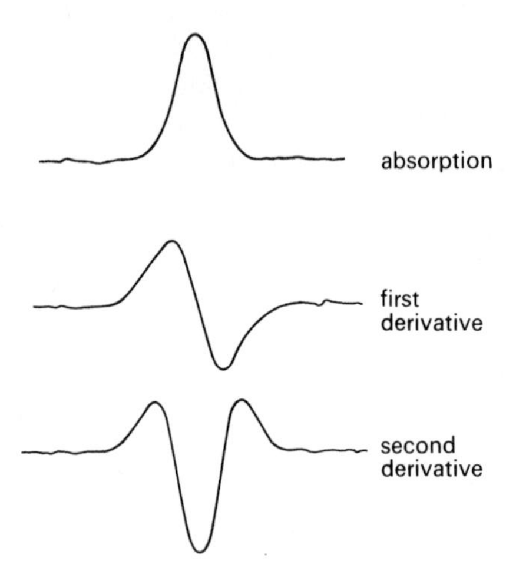

Fig. 4.6 Different ways of presenting the same ESR signal.

Mass spectrometry and related techniques A mass spectrometer is an instrument in which ions are produced, usually by electron bombardment, and then accelerated under the influence of a magnetic field towards a detector. The ions are characterized by the ratio of their mass to their charge. The mass spectrometer may be regarded as an instrument

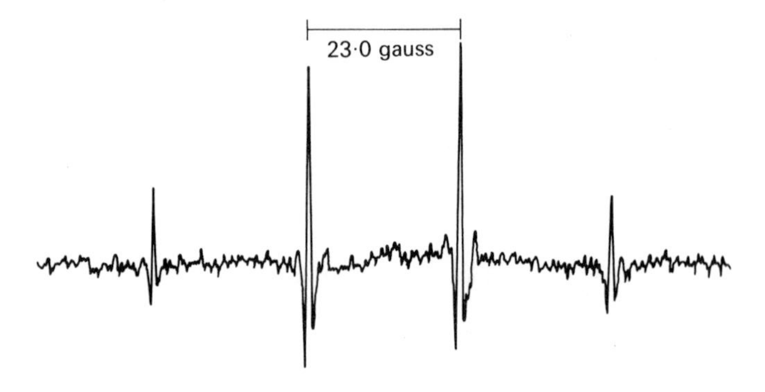

Fig. 4.7 ESR spectrum of liquid methane during irradiation at −176°C.

for performing radiation chemistry in a specialized environment which offers unambiguous characterization of some of the intermediates produced.

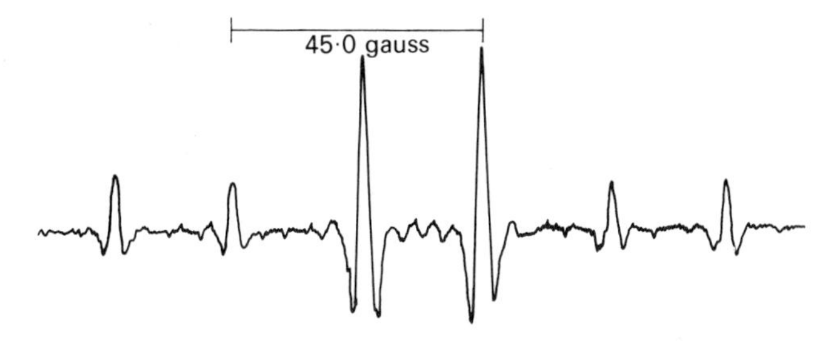

Fig. 4.8 ESR spectrum of solid cyclohexane during irradiation at −10°C.

Modern mass spectrometers often use the Dempster principle [24]. The ions may be produced by allowing a beam of electrons with energies of up to about 50 eV to impinge on the gas being studied. Alternatively α-particles or other sources may be used to ionize the gas. The ions are accelerated by a potential *V*, thus acquiring an energy *Ve*, where e is the charge of the particle.

$$Ve = \tfrac{1}{2}mv^2 \tag{4.32}$$

where *m* is the mass of the particle and v the velocity. After being accelerated, a narrow beam of the particles passes into a magnetic field between two plates. This causes the particles to move along a circular path, after which they go through a slit to a collector. The radius of curvature *r* of the path is determined by the equation:

$$e/m = v/Hr \tag{4.33}$$

where *H* is the magnetic field. Combining 4.32 and 4.33 we obtain:

$$e/m = 2V/H^2r^2 \tag{4.34}$$

If *H* is kept constant, particles with different values of *e/m* can be caused to pass into the detector by altering the value of *V*. Alternatively *V* can be kept constant and *H* altered. Hence a graph of intensity against *V* or *H* with the other constant will give a series of peaks corresponding to ions with different values of *e/m*. The ions observed are those resulting from unimolecular fission of the excited ions produced by the electron beam. The metastable ions which decompose after leaving the ion source and before entering the collector will show up as non-integral peaks. The determination of the fragmentation pattern as such is the basic role of the mass spectrometer. The fragmentation patterns are characteristic of the molecules bombarded, so that mass spectrometers can also be used for quantitative analysis of gases and volatile liquids and solids, and in this respect find a use in radiation chemistry as in other fields where analysis is required. The vast majority of mass spectrometric studies are made with positive ions, which cannot fail to appear when molecules are bombarded with sufficiently energetic electrons, but some studies have been made with negative ions, where they can be produced.

As well as being used to obtain unimolecular fragmentation patterns, mass spectrometers have also been used to help to obtain some of the interlocking thermochemical data (including much of that in Tables 4.1 and 4.4–4.7) which is basic to radiation chemistry as well as other fields. Heats of formation of ions and radicals for example may be derived from appearance potentials. The appearance potential of an ion may be defined as the least energy required to produce the ion and accompanying neutral fragment or fragments from a molecule. The appearance potential depends on the reaction producing the ion. For example, if we assume the products to have no excess excitation energy, the appearance potential of the ion R_1^+ in the reaction:

$$R_1R_2 + e^- \longrightarrow R_1^+ + R_2\cdot + 2e^- \tag{4.35}$$

which is the same as the heat of the reaction, is given by:

$$A(R_1^+) = \Delta H = \Delta Hf(R_1^+) + \Delta Hf(R_2\cdot) - \Delta Hf(R_1R_2) \tag{4.36}$$

where the quantities on the right hand side are the heats of formation of the various entities involved. The appearance potential of the ethyl ion in the mass spectrum of ethane is 12.8 eV and in the mass spectrum of propane it is 12.2 eV [25]. From standard tables (see Table 4.4) the heat of formation of atomic hydrogen is 2.26 eV, of ethane -0.88 eV and of propane -1.08 eV. Hence we have:

$$12.8 = \Delta Hf(C_2H_5^+) + 2.26 + 0.88 \tag{4.37}$$

$$12.2 = \Delta Hf(C_2H_5^+) + \Delta Hf(CH_3\cdot) + 1.08 \tag{4.38}$$

from which $\Delta Hf(C_2H_5^+) = 9.7$ eV which may be compared with 9.5 eV from the sum of the heat of formation of the ethyl radical (Table 4.4) and the ionization potential of ethyl (Table 4.1). $\Delta Hf(CH_3\cdot)$ becomes 1.5 eV (cf. 1.44 in Table 4.4).

Another important relationship is:

$$A(R_1^+) = I(R_1\cdot) + D(R_1 - R_2) \tag{4.39}$$

where $I(R_1\cdot)$ is the ionization potential of the radical $R_1\cdot$. Values for bond dissociation energies, $D(R_1—R_2)$, can be obtained mass spectrometrically if values of $I(R_1\cdot)$ are known. Bond dissociation energies can also often be obtained from appearance potentials alone. For example, the calculation above gave the heat of formation of $CH_3\cdot$ as 1.5 eV, so that the bond dissociation energy of $CH_3—CH_3$ would be $0.88 + 2 \times 1.5 = 3.9$ eV.

Mass spectrometers are most important tools for the investigation of ion-molecule reactions. In fact although ion-molecule interactions had been considered as early as 1905, it was not until they began to be studied by mass spectrometry in 1952 that they could be adequately investigated [26]. Since 1952, numerous investigations have been made using mass spectrometers and related techniques to investigate ion-molecule reactions in the gas phase [16].

The principal evidence that an ion-molecule reaction is occurring in an ordinary mass spectrometer is obtained from measurements of the dependence of ion intensity on pressure. When an ion-molecule reaction is occurring, the intensity of the product ions (secondary ions) is proportional to the square of the pressure. The intensity of primary ions is proportional to the first power of the pressure. Hence the peak height for the secondary ion divided by the peak height for some primary ion should be directly proportional to pressure. The slope of the pressure dependence can be used to measure the reaction cross-section (or rate constant). The dependence of the ion intensity on the electric field in the ionization chamber is another important parameter for investigation since in theory the cross-section for the reaction should be proportional to the inverse of the velocity of the accelerated ion (cf. Equation 4.7) that is, to the inverse square root of the potential V. In fact, however, this relationship is not always observed, and the pressure-dependence method is the most reliable way of establishing that a given ion has its origin in an ion-molecule reaction.

Having established that an ion originates in an ion-molecule reaction, it is necessary to establish the nature of the ion which has given rise to it. The neutral reactant is of course the gas which has been introduced into the instrument. The nature of the reactant ion can often be established by comparing the intensity of the secondary ion with that of various suspected primary ions as a function of ionization energy, since secondary ions cannot be formed at energies which are less than the appearance potential of the primary ions which have given rise to them. Once the nature of the reactant ion, the neutral reactant and the product ion have been established, the nature of the neutral product is obtained by inference.

Numerous variants of the simple mass spectrometric method can be employed to study ion-molecule reactions, including instruments combining two mass spectrometers, various beam techniques and ion-cyclotron resonance [27].

Pulse radiolysis As explained in Chapter 1, certain electron accelerators can deliver their radiation in intense single pulses each lasting typically around one microsecond or less. A single one of these pulses can produce concentrations of intermediates which are high enough to be studied by methods such as ultraviolet or visible absorption spectroscopy, or electrical conductivity. Apparatus based on this idea was first brought into operation in about 1960. The method is called pulse radiolysis, and is the radiation-chemical analogue of flash photolysis. It has proved to be a very powerful tool for the study of intermediates [28].

A schematic diagram of typical pulse radiolysis apparatus is shown in Fig. 4.9. The lamp provides a continuous beam of light which passes through

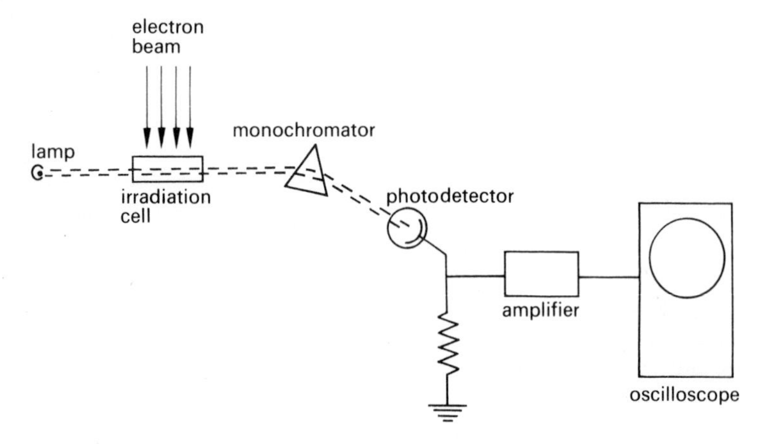

Fig. 4.9 Layout of pulse radiolysis equipment for kinetic spectrophotometry.

the irradiation cell. The monochromator is used to select wavelengths at which the intermediates absorb, and the photodetector converts the light signal into an electrical signal so that changes in light intensity can be followed. The accelerator is used to deliver a single pulse of radiation to the sample. The transmission of the sample before and after the pulse is then displayed as a function of time on the oscilloscope screen. A typical oscilloscope trace is shown on Fig. 4.10. The vertical drop in the figure corresponds to the formation of absorbing species by the radiation and the curve corresponds to the disappearance of species by reaction.†

† Oscilloscope traces can also be presented the other way up so that a *rise* corresponds to the formation of absorbing species, and a *drop* to an increase in transmission.

A variation of the method replaces the steady beam of light by a flash of light which is triggered to go off at a set time after the radiation pulse: instead of detection via a monochromator and a photocell, a spectrograph and a photographic plate may then be used to record the spectrum of those intermediates which are present at the time when the flash of light goes through the cell. Another variant of the method measures

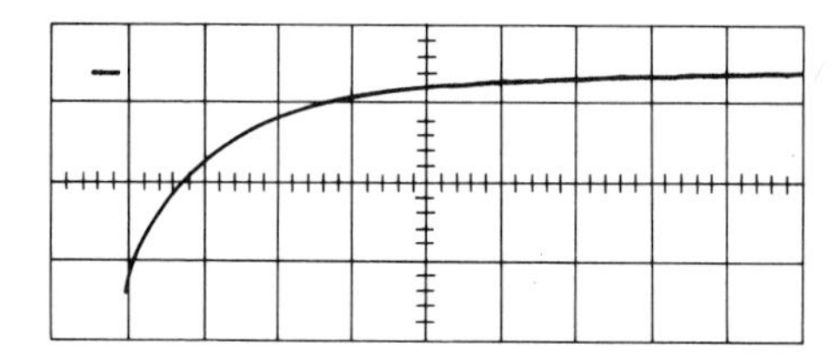

Fig. 4.10 Typical pulse radiolysis oscilloscope trace. The vertical scale corresponds to percentage transmission and the horizontal scale to time. In most experiments one horizontal division would be somewhere between 10^{-7} and 10^{-2} s.

transient changes in conductivity following the pulse. The method can be used to study processes taking place in times as short as about 20 picoseconds (0.02 ns) by making use of the fine-structure pulses of which each single pulse from a linear accelerator is composed [29]. In this method, Cerenkov light produced by the electron beam itself is used as the analysing light. The light's time of arrival at the irradiation cell is varied by using mirrors to cause it to travel various distances (speed of light = 30 cm per ns).

To measure the absorption spectrum of a transient species from oscilloscope traces like that of Fig. 4.10 it is first necessary to measure at a series of wavelengths the drop in transmission caused by a pulse of radiation. Absorbance is then calculated from the drop in transmission. Now absorbance (optical density), *OD*, is related to the molar extinction coefficient, ε in units of $\text{M}^{-1}\ \text{cm}^{-1}$, by the relationship:

$$OD = \varepsilon C l \qquad (4.40)$$

where C is the molar concentration of the species and l the cell path length in centimetres. The concentration of the species can be calculated from the absorbed dose if the G-value is known. Hence ε can be obtained at a series of wavelengths. One of the earliest achievements of pulse radiolysis was to measure the spectrum of the hydrated electron in such a way (Fig. 4.11) [30]. The treatment discussed here assumes that the drop in transmission is due exclusively to the formation of transient species. If the situation is otherwise, for example if the radiation causes the immediate disappearance of a solute which absorbs light, then appropriate modification is necessary.

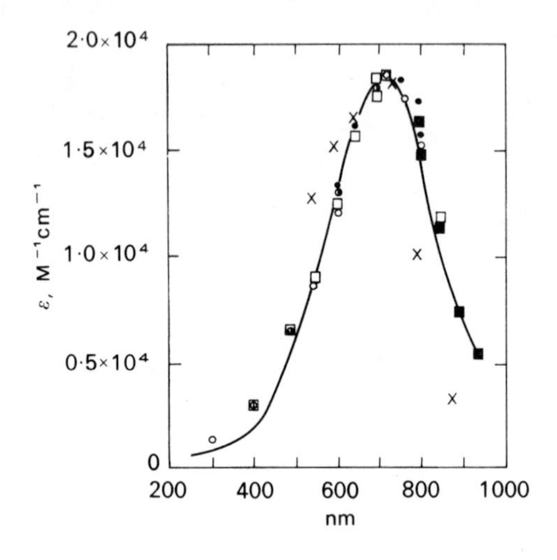

Fig. 4.11 Absorption spectrum of the hydrated electron. The various symbols show experimental points obtained by different workers.

The reactions of transient species can be studied quantitatively by analysing the part of the oscilloscope trace which corresponds to disappearance of the species. One of the simplest cases is where an absorbing species, *R*, produced from a solvent is disappearing by reaction with a solute, *S*, present in great excess. In such a case the concentration of the absorbing species will obey the law:

$$\frac{-d[R]}{dt} = k_{II}[R][S] \tag{4.41}$$

where k_{II} is the second order rate constant for the reaction. Hence:

$$\ln [R]/[R_0] = -k_{II}[S]t \tag{4.42}$$

and:

$$\ln OD/OD_0 = -k_{II}[S]t \tag{4.43}$$

where OD and OD_0 are the absorbance at time t and the initial absorbance respectively. If log absorbance is plotted against time, a straight line will be obtained and k_{II} can be calculated from the slope.

Another elementary case is a simple second order reaction. If a species *A* is reacting with a species *B*, we have:

$$\frac{-d[A]}{dt} = \frac{-d[B]}{dt} = k_{II}[A][B] \tag{4.44}$$

If *A* and *B* are present at equal concentrations, the right hand side may be written $k_{II}[A]^2$, and we have:

$$\frac{1}{[A]} = \frac{1}{[B]} = k_{II}t + \frac{1}{[A_0]} \tag{4.45}$$

In a simple case *A* may absorb light of the wavelength being used, but *B* and the reaction products may not, so that we would have:

$$\frac{1}{OD} = \frac{k_{II}t}{\varepsilon l} + \frac{1}{OD_0} \tag{4.46}$$

where *OD* is the absorbance at time t, ε is the extinction coefficient of *A*, and l is the path length. In such a case a plot of $1/OD$ against t would be linear, and k_{II} could be determined from the slope.

If the species *A* is reacting with another species *A*, the rate of loss of *A* will be twice as great (for equal collision efficiency) as if *A* were reacting with *B*. We may therefore write:

$$\frac{-d[A]}{dt} = 2k_{II}[A] \tag{4.47}$$

and:

$$\frac{1}{[A]} = 2k_{II}t + \frac{1}{[A_0]} \tag{4.48}$$

These proportionalities may of course also be expressed in the forms:

$$\frac{-d[A]}{dt} = k_{II}[A]^2 \tag{4.49}$$

and:

$$\frac{1}{[A]} = k_{II}t + \frac{1}{[A_0]} \tag{4.50}$$

where the value of the rate constant defined in this way is twice that defined by Equation 4.48. Confusion can arise if the rate constants for reactions of identical species are not carefully defined.

In practice the kinetics of the observed changes in absorption are rarely as simple as those discussed above, except in carefully chosen systems, but a great deal of information about postulated reactions can be obtained by attempting to fit a proposed mechanism to the absorption changes observed under various experimental conditions.

PROBLEMS

1. What would be the *G*-value for primary production of the ^1E_1u (6.9 eV) level of benzene if the oscillator strength for the transition is 0.9 and the distribution of primary activations is that given by the optical approximation? Take the total oscillator strength for benzene to be equal to the number of valence electrons in the molecule.

[Ans: 0.43]

2. If the irradiation of an aromatic solution produces solvent triplets in a yield, independent of solute concentration, of $G = 4.0$, with a mean life in the pure solvent of 1.4×10^{-7} s, what would be the yield of solute triplets if the solvent triplets could transfer energy to solute with $k = 10^{10}$ M^{-1} s^{-1} (*a*) for a 1.25×10^{-4}M solution, (*b*) for a 10^{-3}M solution?

[Ans: (*a*) $G = 0.6$, (*b*) $G = 2.3$]

3. If the rate of reaction between hydrated electrons (radius 0.27 nm, diffusion constant 4.8×10^{-5} cm^2 s^{-1}) and $Fe(CN)_6^{3-}$ (radius 0.28 nm diffusion constant 9×10^{-6} cm^2 s^{-1}) is governed by diffusion, what value would be expected for the rate constant of the reaction at 27°C?

[Ans: 2.0×10^9 M^{-1} s^{-1}]

4. A solution of iodine in cyclohexane (1×10^{-4}M) is irradiated with γ-rays at a dose-rate of 5×10^{15} eV g^{-1} s^{-1}. Assuming cyclohexane gives 5 radicals per 100 eV, calculate the concentration of iodine after one minute (density of cyclohexane = 0.78 g cm^{-3}).

[Ans: 0.9×10^{-4}M]

5. A liquid of density of 0.8 g cm^{-3} containing dissolved oxygen at a concentration of 10^{-3}M is irradiated with γ-rays at a dose-rate of 10 000 rads per minute. Free radicals are produced in the bulk of the solution with $G = 6$ and disappear by reaction with oxygen, with $k = 10^9$ M^{-1} s^{-1}. (*a*) Calculate the steady state concentration of radicals. (*b*) What fraction of the radicals will react with each other if the rate constant defined by $-d[R\cdot]/dt = k[R\cdot]^2$ is 10^9 $\mathrm{M^{-1}\,s^{-1}}$?

[Ans: (*a*) 8.3×10^{-13}M (*b*) 1.2×10^{-11}]

6. The appearance potential of CH_3^+ in the mass spectrum of methane is 14.3 eV. Using data from Table 4.4, calculate the heat of formation of CH_3^+ in kcal per mole. What appearance potential in eV would be expected for CH_3^+ in the mass spectrum of bromomethane if the heat of formation of bromomethane is -8.4 kcal per mole and of atomic bromine 26.741 kcal per mole?

[Ans: 260, 12.8]

7. A pulse of radiation is given to a 5×10^{-2}M aqueous solution of glycine. Absorbances at 600 nm at 0, 1, 2, 3, 4, 5 μs after the pulse are, respectively, 4.8, 3.15, 2.05, 1.34, 0.88, 0.57×10^{-3}. (*a*) Calculate the rate constant for the reaction between hydrated electrons and glycine. (*b*) What is the half-life of hydrated electrons in this solution? If the pulse had produced 10^{-5}M hydrated electrons, what would have been the initial rates of reaction of hydrated electrons (*c*) with glycine (*d*) with each other, if the rate constant for the mutual reaction of hydrated electrons, defined by $-d[e^-_{aq}]/dt = k[e^-_{aq}]^2$, is 1×10^{10} $\mathrm{M^{-1}\,s^{-1}}$?

[Ans: (*a*) 8.5×10^6 $\mathrm{M^{-1}\,s^{-1}}$, (*b*) 1.6 μs, (*c*) 4.25 $\mathrm{M\,s^{-1}}$, (*d*) 1 $\mathrm{M\,s^{-1}}$]

REFERENCES

1. R. L. Platzman. *The Vortex*, **28**, 372–85 (1962), 'Superexcited states of molecules, and the primary action of ionizing radiation'.
2. R. L. Platzman, in *Radiation Research*, ed. G. Silini. North-Holland, Amsterdam, 1967, pp. 20–42, 'Energy spectrum of primary activations in the action of ionizing radiation'.
3. R. L. Platzman. *Radiat. Res.*, **17**, 419–25 (1962), 'Superexcited states of molecules'.
4. Based on J. L. Franklin, J. G. Dillard, H. M. Rosenstock, J. T. Herron, K. Draxl and F. H. Field. *'Ionization Potentials, Appearance Potentials, and Heats of Formation of Gaseous Positive Ions'*, *NSRDS-NBS* 26, Washington, D.C., 1969.
5. R. Voltz, in *Progress and Problems in Contemporary Radiation Chemistry. I*, ed. J. Teplý. Institute of Nuclear Research of the Czechoslovak Academy of Sciences, Prague, 1971, pp. 139–85, 'Comments on the activated species in irradiated organic media'.
6. For example, J. G. Calvert and J. N. Pitts jr. *Photochemistry*. Wiley, New York, N.Y., 1966.

7. R. Voltz, in *Actions Chimiques et Biologiques des Radiations,* treizième série, ed. M. Haïssinsky. Masson, Paris, 1969, pp. 1–55, 'Radioluminescence des milieux organiques'.
8. J. B. Birks. *The Theory and Practice of Scintillation Counting.* Pergamon Press, New York, N.Y., 1964.
9. W. P. Jesse and J. Sadauskis. *Phys. Rev.,* **100**, 1755–62 (1955), 'Ionization by alpha particles in mixtures of gases'.
10. O. Stern and M. Volmer. *Physik. Z.,* **20**, 183–8 (1919), 'Über die Abklingungszeit der Fluoreszenz'.
11. P. Debye. *Trans. Electrochem. Soc.,* **82**, 265–71 (1942), 'Reaction rates in ionic solutions'.
12. H. M. Rosenstock, M. B. Wallenstein, A. L. Wahrhaftig and H. Eyring. *Proc. Nat. Acad. Sci. U.S.,* **38**, 667–78 (1952), 'Absolute rate theory for isolated systems and the mass spectra of polyatomic molecules'.
13. American Petroleum Institute, Research Project 44. *Catalog of Mass Spectral Data.* Carnegie Institute of Technology, Pittsburgh, Pa.
14. V. I. Vedeneyev, L. V. Gurvich, V. N. Kondrat'yev, V. A. Medvedev and Ye. L. Frankevich. *Bond Energies Ionization Potentials and Electron Affinities.* Edward Arnold, London, 1966.
15. G. R. Freeman. *Radiat. Res. Rev.,* **1**, 1–74 (1968), 'The radiolysis of aliphatic and alicyclic hydrocarbons'.
16. J. H. Futrell and T. O. Tiernan, in *Fundamental Processes in Radiation Chemistry,* ed. P. Ausloos. Interscience, New York, N.Y., 1968, pp. 171–280, 'Ion-molecule reactions'.
17. P. Ausloos. *Progr. Reaction Kinetics,* **5**, 113–79 (1970), 'Ion-molecule reactions in radiolysis and photoionization of hydrocarbons'.
18. W. F. Schmidt and A. O. Allen. *J. Phys. Chem.,* **72**, 3730–36 (1968), 'Yield of free ions in irradiated liquids; determination by a clearing field.'
19. Based on I. T. Myers, in *Radiation Dosimetry,* 2nd Edn, I, eds F. H. Attix and W. C. Roesch. Academic Press, New York, N.Y., 1968, pp. 317–30, 'Ionization'.
20. Most of these values are taken from R. A. Holroyd, in *Fundamental Processes in Radiation Chemistry,* ed. P. Ausloos. Interscience, New York, N.Y., 1968, pp. 413–514, 'Organic liquids'.
21. A. Kuppermann, in *Actions Chimiques et Biologiques des Radiations,* cinquième série, ed. M. Haïssinsky. Masson, Paris, 1961, pp. 85–166, 'Diffusion kinetics in radiation chemistry'.
22. H. A. Schwarz. *J. Phys. Chem.,* **73**, 1928–37 (1969), 'Applications of the spur diffusion model to the radiation chemistry of aqueous solutions'.
23. R. W. Fessenden and R. H. Schuler. *J. Chem. Phys.,* **39**, 2147–95 (1963), 'Electron spin resonance studies of transient alkyl radicals'.
24. A. J. Dempster. *Phys. Rev.,* **11**, 316–25 (1918), 'A new method of positive ray analysis'.
25. F. H. Dorman. *J. Chem. Phys.,* **43**, 3507–12 (1965), 'Second differential ionization-efficiency curves for fragment ions by electron impact'.
26. В. Л. Тальрозе и А. К. Любимова. *Докл. Акад. Наук СССР,* **86**, 909–12 (1952), 'Вторичные процессы в ионном источнике масс-спектрометра'.
27. E. W. McDaniel, V. Čermák, A. Dalgarno, E. E. Ferguson and L. Friedman. *Ion-Molecule Reactions.* Wiley-Interscience, New York, N.Y., 1970.

28. M. S. Matheson and L. M. Dorfman. *Pulse Radiolysis.* M.I.T. Press, Cambridge, Mass., 1969.
29. M. J. Bronskill and J. W. Hunt. *J. Phys. Chem.*, **72**, 3762–66 (1968), 'A pulse-radiolysis system for the observation of short-lived transients'.
30. E. J. Hart and M. Anbar. *The Hydrated Electron.* Wiley-Interscience, New York, N.Y., 1970, p. 42.

Inorganic solids

The effects of elastic collisions and, in certain cases, nuclear transformations, show up most clearly in the irradiation of inorganic solids such as metals where excitation and ionization are without significant effect and where structural defects play a key role in determining the properties of the material [1]. Excitation and ionization can cause effects in the irradiation of other inorganic solids such as semiconductors and ionic crystals but, here too, structural defects play an important part in the interpretation of the effects observed.

In the interpretation of effects produced by displacement of atoms, it is an important general feature that the various kinds of defect can move through the solid, the more rapidly the higher the temperature, forming clusters of defects of various sizes or becoming annihilated by recombination of interstitials with vacancies. Interstitial atoms in metals may begin to move at temperatures as low as about 10°K. Vacancies migrate at higher temperatures.† Accordingly the effect of irradiation is a function of temperature. Effects can be modified by heating after irradiation and in some cases, providing no permanent damage has been introduced by nuclear transformations, can even be annealed out altogether.

The direct study of the short-lived intermediates produced by excitation and ionization (pp. 83–94) finds a counterpart in this field in direct observations of the structural defects responsible for changes in physical properties. Electron spin resonance itself has been widely used with non-

† When a vacancy is present in a regular crystalline lattice, the movement of a neighbouring atom into the vacancy results in the vacancy having migrated one spacing.

metals. Defect clusters as small as about 1 nm can be seen visually by means of electron microscopy of thin foils. The position of individual atoms can be studied by means of field ion microscopy [2]. In this technique the sample is a needle point facing a phosphor screen. An electrostatic field is applied between the sample and the screen while helium or neon are allowed to become adsorbed on the sample. The atoms become ionized at points where the field is locally enhanced and then travel in straight lines to the screen, where they produce an image which reflects the pattern of atoms on the sample surface. The surface atoms can be evaporated, leaving the next layer of atoms exposed, and in this way the whole three-dimensional structure can be examined. Typical pictures, showing vacancies in a crystal of platinum [2], are shown in Fig. 5.1.

The study of radiation effects in inorganic solids may be regarded as a branch of solid state physics, to which subject it has made substantial contributions. It is of vital importance for nuclear technology and is also relevant to several other fields. It meets up with other parts of radiation chemistry at several points, for instance in the effect of radiation on glasses and ionic crystals and in the effect of radiation on reactions occurring at surfaces. Some of the concepts in this field may apply to the radiolysis of substances like polyphenyls, which are resistant to excitation and ionization but could be damaged by the displacement of atoms through elastic collisions.

Stored energy

When an atom is displaced from its normal lattice site by irradiation it often becomes trapped at another, less stable, position. Under the influence of thermal vibrations, atoms are able to return to more stable positions, liberating stored energy. Such an effect was first seen in 1815, in which year Berzelius reported that a marked evolution of energy took place when the mineral gadolinite was subjected to heat. Later it was found that the gadolinite contained radioactive elements and that some of the energy arising from radioactive decay had become stored in the mineral, to be released again on heating. Some minerals contain as much as 65 calories of stored energy per gram ($270\ J\ g^{-1}$) [3].

In the case of metals, some of the radiation-produced defects can recombine extremely easily, and it is necessary to irradiate at very low temperatures in order to store appreciable energy. In one typical experiment, copper foils (0.13 mm thick) immersed in liquid helium (4°K) were first irradiated with 1.2 MeV electrons to a fluence of 9×10^{17} electrons per cm^2, and then the irradiated and unirradiated specimens were warmed up by heating at the same rate [4]. Up to about 50°K, the temperature of the irradiated specimen was always slightly higher than that of the unirradiated because of the release of stored energy. The amount of heat liberated is shown in Fig. 5.2 as a function of temperature. Changes taking place on warming to about 60°K are called 'stage I annealing'. Further energy evolution occurs at other stages in the heating process but most of the energy is liberated below room temperature.

The large liberation of energy in stage I annealing is attributed to the movement of some of the interstitials into nearby vacancies and to the

Fig. 5.1 Field ion microscopy of crystal of platinum. Pictures 2, 3 and 4 show vacancies which were revealed after successive layers had been removed by field evaporation.

movement of other interstitials into more distant impurity traps, grain boundaries and so on. Recombination of each Frenkel defect in copper liberates about 5 eV of energy. Comparable amounts of energy are liberated for other metals. The liberation of energy at higher temperatures (stages II, III and so on) is due to residual interstitials moving from shallow into deeper traps, and to various aggregations and recombinations.

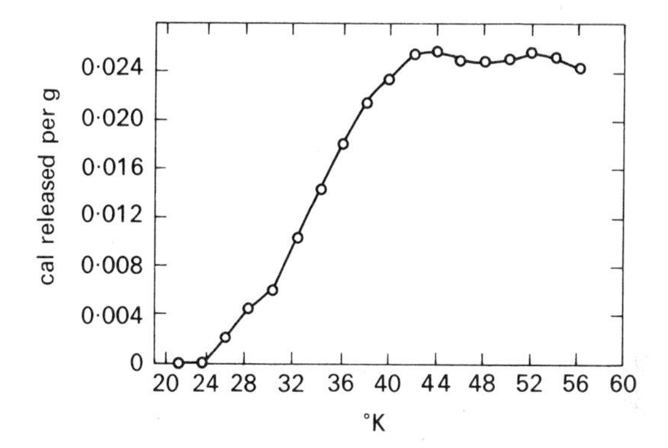

Fig. 5.2 Total energy release on allowing an irradiated sample of copper to warm up to various temperatures.

At room temperature and above, very much larger amounts of energy can be stored in certain non-metals than in metals. Graphite which has been irradiated in a reactor to more than 5×10^{20} *nvt* (total neutrons)† still retains appreciable energy even after heating to 1000°C, although when small doses are used ($<10^{20}$ *nvt* (total)) most of the energy can be liberated by heating to 300°C [5]. Stored energy can be of great practical importance in the operation of graphite-moderated nuclear reactors working at relatively low temperatures, because of the possibility that the energy stored in the graphite may eventually be released catastrophically. Such an accident happened at Windscale, England, on 10 October 1957. Excess heating due to the release of stored energy caused a reactor to catch fire, releasing tens of thousands of curies of ^{131}I and other active species into the air. Fortunately there is little or no possibility of such accidents with graphite-moderated reactors working at higher temperatures.

Trapping of atoms at metastable positions is not the only way in which radiation energy can be stored in a material, neither is heat the only

† As noted in Chapter 3 it is very difficult to measure the energy distribution of the fast neutrons in a nuclear reactor, so that doses quoted have often been somewhat rough, especially in earlier work.

form in which it can be liberated again. For example, the energy level of a material containing trapped electrons and positive holes is higher than that of the original material, and energy may be released again in the form of light (p. 112). Also radiation energy often appears as chemical energy when labile chemicals are irradiated. Irradiation can also *liberate* stored energy. In every case the effect of radiation is to be understood in mechanistic terms, and the gross thermodynamics of the system provides little guidance as to what is happening.

Electrical conductivity

Metals Neither radiation-produced excitation nor ionization can produce an effect on the conductivity of metals because so many conduction electrons are already present. However, defects produced by elastic collisions can decrease conductivity by scattering the conduction electrons, so interfering with their ordered flow in an electric current. The greatest effects are noted at low temperatures, partly because of the tendency of defects to anneal out under the influence of lattice vibrations and partly because lattice vibrations themselves scatter conduction electrons and so tend to mask effects due to defects.

In one of the pioneering studies, copper, silver and gold wires (0.13 mm thick) were irradiated with 12 MeV deuterons (range in copper 0.20 mm) at temperatures near 10°K, and the increase in resistivity was measured at the same temperature [6]. The increase in resistivity with fluence is shown in Fig. 5.3. In the case of copper, which is the simplest and best understood of the metals, the resistivity increase associated with 1 per

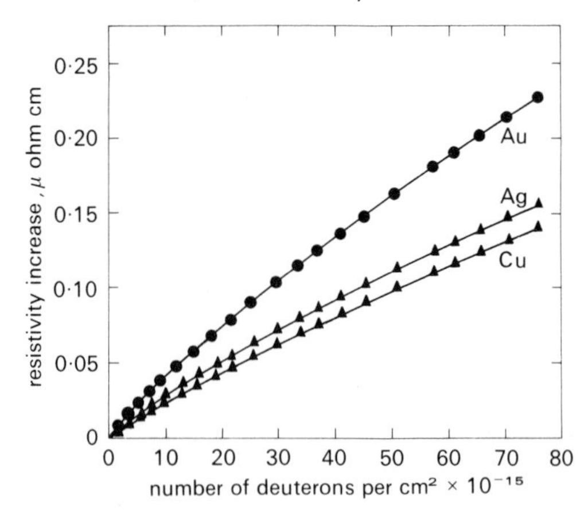

Fig. 5.3 Increase in resistivity of copper, silver and gold as a function of deuteron fluence.

cent of the atoms being displaced to form Frenkel defects is believed from a variety of theoretical and experimental considerations to be around 2.5 μ ohm cm. About half of this is associated with the interstitial and half with the vacancy. Hence in the experiments of Fig. 5.3 the radiation had initially displaced 0.009 per cent of the atoms per 10^{16} deuterons per cm^2. The rate of production of defects had decreased as the irradiation proceeded.

The annealing of resistivity changes is similar to that of stored energy, since similar defects are responsible for both. In the experiments of Reference 4, about 40–50 per cent of the resistivity increase in copper disappeared below 60°K, and 92 per cent by 300°K. The annealing of irradiated silver and gold is qualitatively similar, 90 per cent recovery being obtained by 300°K in both cases.

Disturbances produced by radiation can change conductivity indirectly by giving rise to other physical changes which themselves change the conductivity. A good example is in the irradiation of order-disorder alloys where large changes in conductivity arise from changes in order (p. 107).

Semiconductors† Excitation and ionization produce much greater effects on the conductivity of semiconductors than on metals, but this effect disappears after the irradiation stops. Changes after irradiation are attributable to the production of various kinds of defect whose principal effects are to act either as acceptors or donors of electrons. If an *n*-type semiconductor is irradiated, the defects may accept conduction electrons so that the conductivity may drop. Further production of electron-accepting defects could enable positive holes to be produced so that the semiconductor would change over to the *p*-type. Continued production of such defects would lead to an increase in *p*-type conductivity. Electron donating defects would increase the conductivity of *n*-type semiconductors and decrease that of *p*-type. Such effects produce very strong effects on the electrical properties of semiconductors, so that the electrical properties of semiconductors are very much more sensitive to radiation than those of metals. Furthermore, scattering effects, as discussed above for metals, are of minor importance in semiconductors except at very low temperatures.

It is evident that the effect of radiation on semiconductors must be

† Most semiconductors owe their conductivity to impurities, which may be donors or acceptors of electrons. When the current is carried predominantly by electrons arising from donors the semiconductor is referred to as *n*-type (for negative). When the current is carried by 'holes' arising from the acceptance of electrons, the semiconductor is referred to as *p*-type (for positive).

exceedingly complicated. Some typical overall observations are listed in Table 5.1.

Table 5.1 *Some effects of radiation on semiconductors* [7]

Material	*Response to irradiation*
n-type Ge	converted to *p*-type
p-type Ge	approaches a limiting hole concentration (7×10^{16} cm^{-3})
n- and *p*-type Si	carrier concentrations approach intrinsic value
n-type InSb	approaches a limiting electron concentration (4×10^{16} cm^{-3})
p-type InSb	converted to *n*-type
n-type GaSb	converted to *p*-type
p-type GaSb	approaches a limiting hole concentration ($\sim 10^{16}$ cm^{-3})
n-type AlSb	approaches a limiting electron concentration
p-type AlSb	converted to *n*-type
n-type InAs	electron concentration appears to increase indefinitely
p-type InAs	converted to *n*-type
n-type InP	electron concentration decreases, no evidence of conversion
p-type CdTe	converts to *n*-type

Insulators While an irradiation is taking place electrons are continually being excited into the conduction band, which in insulators contain very few electrons. Hence the resistivity of insulators (including organic insulators) decreases markedly during the irradiation itself. Decreases in resistivity of 100–1000 times at dose rates as low as 8 R per minute are common [8]. After irradiation the resistivity rises gradually to values which are somewhat lower than the initial value because of the formation of defects of various kinds. The decreases in resistivity are not nearly enough to turn the insulators into conductors, and large electrical charges can build up inside insulators subjected to charged particle bombardment. Eventually the charges may be released catastrophically.

Thermal conductivity

Defects produced by radiation will scatter thermal waves just as they scatter conduction electrons. Hence thermal conductivity drops on irradiation. Effects in metals are of comparatively little theoretical or practical importance, but large effects occur in ceramics and in graphite, and these are of considerable technological importance. In the case of graphite a dose of 10^{20} *nvt* (total) at room temperature decreases the thermal conductivity by a factor of about ten [5]. Effects, as always, are less at higher temperatures.

Dimensional changes

The presence of displaced atoms increases the volume of irradiated solids. In the case of metals, appreciable increases in volume due to displaced atoms can be obtained at very low temperatures. Most such defects anneal out at room temperature or above. However, atoms which have been displaced by radiation, including atoms displaced transiently in spikes, do not necessarily return to their original sites. With anisotropic crystals this enables the shape to change on irradiation at ordinary temperatures, even though the volume shows little change. Uranium provides an important example of this behaviour [9]. Single crystals of α-uranium increase in length along the [010] direction (for example, for 0.01 per cent burnup, that is, 0.01 per cent of uranium atoms fissioned, the length increases by 4.2 per cent), and decreases along the [100] direction to the same extent. There is no change along the [001] direction. In polycrystalline specimens a dimensional change still occurs, the extent depending on the method of manufacture, and, in addition, the growth of individual grains causes some wrinkling of the surface (Fig. 5.4) [10].

Although most of the defects in irradiated metals anneal out below room temperature, the partly annealed configurations of defects which are present at higher temperatures, although unable to produce much change in properties such as stored energy content and electrical resistivity, are still capable of producing important changes in dimensions.

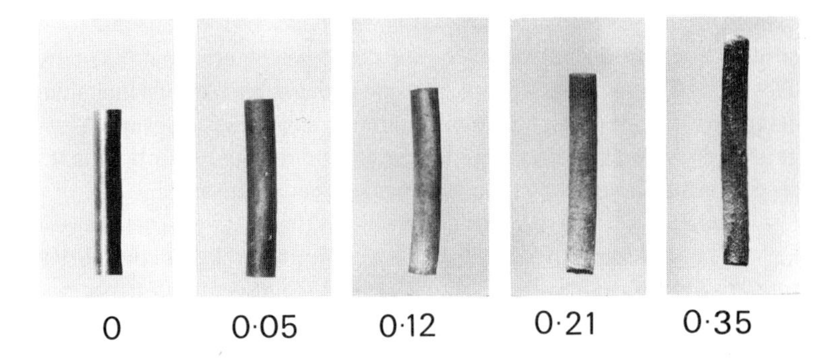

Fig. 5.4 Typical specimens of uranium at various per cent burnup.

This becomes evident in metals irradiated to very high doses (10^{22}–10^{23} *nvt* (fast)) at very high temperatures (say 500–600°C) where vacancies can aggregate on to small amounts of gas to form voids, leaving the interstitials to expand the metal [11]. The possibility of this process occurring in fuel-element cladding and structural materials is a highly important consideration in the design of fast breeder reactors [12]. Different materials exhibit the effect to different extents. For a given metal the effect can often be reduced by prior mechanical treatment so as to increase the number of defects, such as dislocation lines, and thus provide sites where vacancies and interstitials can recombine.

Since displacements in covalent and ionic solids do not anneal as readily as in metals, large changes in dimensions can be obtained with these materials at ordinary temperatures. The expansion of graphite is of special interest with regard to reactor design. In this case expansion occurs through the forcing apart of the lattice planes by interstitial atoms. There is a slight contraction in the lattice plane itself. Hence the expansion is anisotropic. The expansion of polycrystalline graphite is less than that of single crystals because of the porous structure but the expansion is still anisotropic, varying according to the mode of manufacture of the specimen. Typically, a dose of 10^{20} *nvt* (total) at 30°C gives a volume increase in polycrystalline graphite of 0.2 per cent [5]. In pyrolytic carbon, whose structure is intermediate between the crystalline and the amorphous, gross shrinkage can occur on irradiation, resulting in void formation. This is an important phenomenon in the use of pyrolytic carbon as a coating round the fuel particles in high temperature gas cooled reactors.

Nuclear transformations as such can give rise to significant changes in dimensions owing to the formation of gases or solid fission products. The reactor irradiation of uranium provides a good example. There is little effect on irradiation below 350–400°C, but a marked swelling can be seen on prolonged irradiation above this temperature. In a typical experiment, a volume increase of about 10 per cent has been seen on irradiation at 500–800°C to 0.25 per cent burnup [13]. Swelling can be reduced by alloying and is much less in uranium oxide and uranium carbide fuels than in uranium metal [14]. Other ductile materials also exhibit swelling. When gases are formed in brittle materials, swelling does not occur but cracks tend to develop instead (cf. poly(methyl methacrylate), p. 215).

Although swelling is less with ceramic fuels than with the metal, swelling and gas release in these materials is a major technological problem in modern fuels for both advanced gas cooled reactors and fast reactors. In this case burnup can exceed 10 per cent and the temperature at the centre of the fuel can exceed 2000°C. Gas can be transported to the grain boundaries by the movement of bubbles where it accumulates and eventually becomes released by cracking. Also large grain boundary bubbles move up the temperature gradient to the fuel centre by a mechanism involving evaporation and condensation and, once again, the gas is released into the fuel can.

Order–disorder changes in alloys

Certain alloys exist at the absolute zero in a fully ordered form, in which each atom occupies a definite lattice site, and at higher temperatures in a disordered form in which sites are occupied at random. The alloy Cu_3Au is typical. At equilibrium at the absolute zero the gold atoms are at the corners of a face-centred cubic cell and the copper atoms are at the centres. Above 388°C the equilibrium distribution is random. However, because of the slowness at which equilibrium is reached, a non-equilibrium degree of order is normally present. Changes in order can be followed readily by measuring electrical resistivity, which increases with disorder. Radiation can either increase or decrease order. If an order-disorder alloy containing more disorder than the equilibrium value is irradiated the order can increase [15]. This is attributable to the production of vacancies which assists diffusion of atoms to take place. If an ordered alloy is irradiated with radiation which can produce many displacements per primary knock-on, disordering takes place [16]. This is to be understood as a consequence of replacement collisions and/or of the high temperature in spikes.

Phase changes

Disturbances produced by radiation often cause phase changes in the direction of thermodynamic equilibrium when materials are irradiated in a metastable condition. The ordering of disordered alloys just discussed is one example. One of the simplest examples with pure materials is in the irradiation of white tin at low temperatures. After irradiation the samples change phase in the same way as samples which have been seeded with grey tin [17]. Similarly certain metastable solid solutions precipitate on heating after irradiation in the same way as if nucleated by ageing.

Phase changes away from thermodynamic equilibrium also occur quite often. Disordering phenomena are perhaps not very surprising, but changes can also occur in the direction of thermodynamically disfavoured ordered phases. For example, monoclinic zirconia can be converted into the cubic phase, normally stable only above 1900°C, by irradiation with fission fragments [18]. The change may be caused by the rapidly quenched high temperature in the fission spikes.

Mechanical properties [19]

The mechanical properties of solids are strongly dependent on imperfections in the crystal lattice. Defects can hinder the relatively easy stress-induced movement of dislocations through the crystal and so lead to increased strength and hardness. With metals large changes are observed at very low temperatures, and even at ordinary temperatures the partially annealed configurations of defects can produce considerable changes in mechanical properties. In fact it is with metals that changes in mechanical properties are most important, both from the theoretical and the practical points of view.

One measure of strength is yield strength.† It is very common to find a doubling in the yield strength of metals after irradiation, and much larger effects have been observed. Samples which have a low initial dislocation density show larger percentage charges on irradiation than samples which have a higher initial dislocation density (and correspondingly higher initial yield strength).

The ultimate tensile strength of metals also increases on irradiation, although often to a smaller extent than yield strength. Elongation to fracture usually decreases on irradiation.

Embrittlement on irradiation is of particular technological importance with certain steels as well as with certain other metals. These are ductile above a certain temperature and brittle below it; on irradiation the ductile-brittle transition temperature increases. At the same time the impact strength decreases. The effect is caused by clusters of defects produced by the radiation. Many investigations have shown that these defects disappear rapidly by diffusion processes at temperatures above about half the melting point in °K. Accordingly the steels recover much of their normal hardness and ductility on heating to such temperatures. However, on testing at still higher temperatures such as may be encountered in a breeder reactor (up to about 800°C) the irradiated steels are found to be brittle once more. This is accounted for by the neutrons having reacted with constituents of the steel to give hydrogen and,

† Yield strength is the stress (load) at which a material exhibits a specified permanent set, often taken to be 0.20 per cent. Yield strength is obtained from the stress-strain diagram by drawing a line which starts on the strain axis at the specified set and is parallel to the initial elastic line: the stress at which this line meets the stress-strain curve is the yield strength.

especially, helium, which then diffuse to the grain boundaries to form bubbles which cause the embrittlement [20].

With several materials of importance for nuclear technology, radiation may produce considerable increases in creep, that is, in extension under constant stress as a function of time. The effect is most important for polycrystalline materials such as uranium and graphite. It arises from the dimensional changes discussed above.

Colour changes [21]

Many non-metals are very sensitive to colouration by radiation, even when displacements are not produced. Colour changes often occur with doses of a few tens of thousands of rads or even less. Colouration effects are seen in Nature in the pleochroic haloes surrounding radioactive inclusions in mica and other materials. The phenomenon is best understood for alkali halides.

Alkali halides In contrast to most of the changes so far discussed, the changes in colour of these materials are attributable principally to excitation and ionization rather than to elastic collisions or nuclear transformations. The colouration of the pure crystals depends on the presence of vacant lattice sites [22]. Electrons ejected by radiation become trapped at negative-ion vacancies, forming '*F*-centres'.† The excitation of electrons in these centres then gives rise to an absorption band, called the *F*-absorption band. Similarly the 'positive holes' left by the ejection of electrons move through the crystal (by electron transfer in the opposite direction) and become trapped at positive-ion vacancies, forming *V*-centres. *F*- and *V*-centres are shown diagrammatically in Fig. 5.5 [23], together with some variants which account for other absorption bands which appear either initially or at various stages in the optical or thermal annealing of the crystals. Impurities such as traces of divalent cations introduce additional trapping centres. A typical absorption spectrum is shown in Fig. 5.6 [24]. Irradiated alkali halides appear yellow, green, blue, violet and so on, according to the nature of the alkali metal and the halogen ion, but, in accordance with the mechanism just discussed, independent of the type of radiation used.

Prolonged irradiation of the alkali halides, even with radiation incapable of causing elastic collisions or nuclear transformations, is found to form many more *F*-centres than can be accounted for by the number of vacancies originally present in the crystal. Numerous mechanisms have

† From the German, Farbe.

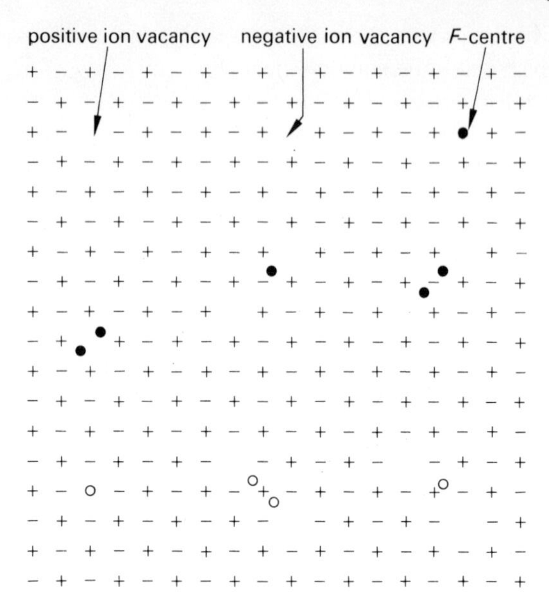

Fig. 5.5 Some colour centres in alkali halides; ● – electron, ○ – hole. The trapped hole may be considered as a site where an electron is missing from one of the negative ions round a positive ion vacancy.

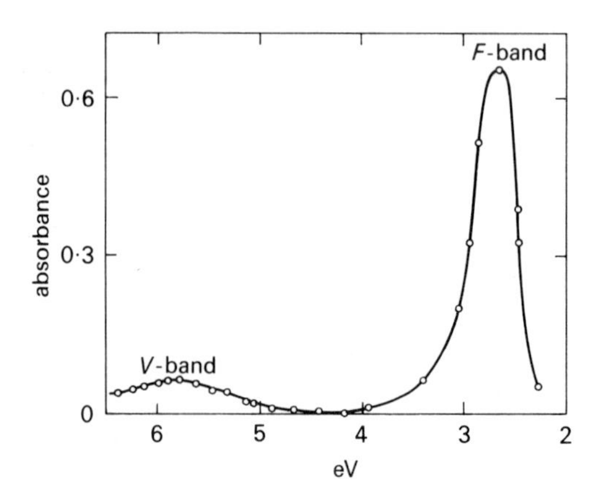

Fig. 5.6 Absorption spectrum of an irradiated sodium chloride crystal.

been proposed to account for this. According to one view, excitons (localized regions of electronic excitation energy) diffuse through the crystal until they encounter sites such as jogs in dislocation lines where the ions are weakly bound [23, 25]. Such sites are shown in Fig. 5.7. Similarly electrons and holes may combine at these sites, liberating energy. The energy becomes converted into lattice vibrational energy, which causes ions neighbouring the dislocation to move into the dislocation, so creating vacancies in the crystal. The vacancy-dislocation units then take up energy from excitons or electron-hole combinations, and the resulting lattice vibrations enable vacancies to escape from dislocations.

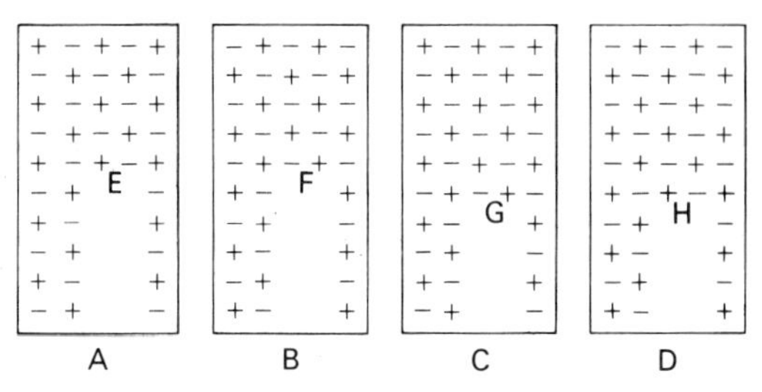

Fig. 5.7 Crystal containing a jog in a dislocation line. The crystal consists of alternate stacks of A and B on top of which are alternate stacks of C and D. The lines E–F and G–H are edge dislocation lines. The intersection F–G is a jog.

According to another mechanism, vacancies are produced through multiple ionization of negative ions [26]. This could happen if inner-shell ionization occurs (p. 26) since loss of an electron would then be followed by an Auger process leading to loss of a second electron. Multiple ionization would convert negative ions into positive ions, whereupon coulombic repulsion would instantly eject them into interstitial positions, leaving behind negative ion vacancies. According to one version of this mechanism, the interstitial ions then capture lost electrons, so as to become neutral. The remaining electrons are then captured by the negative ion vacancies to give *F*-centres. It may be that both this and the dislocation mechanism can operate, their relative importance being a function of temperature [27].

When alkali halides which have been coloured by irradiation are illuminated with light, the vacancies tend to join together into clusters and the colour partly fades owing to recombination of electrons and holes. Mild heating of irradiated alkali halides causes vacancies to cluster whilst strong heating causes defects to disappear and the colour to disappear altogether. The disappearance of colour is accompanied by the emission of light. Light emission from suitable crystals is widely used as a means of measuring radiation doses (p. 53).

Glass and quartz Examples of the colouration of these materials are seen by every research worker who uses glass apparatus for irradiation.

The colouration which develops at low doses is due to the trapping of electrons and holes at impurities. The displacement of atoms introduces extra defects which give rise to additional absorption bands. The colours fade with time, the process being accelerated by heat or light.

The colouration of glass can sometimes be a real disadvantage in practice, for example when it is desired to have windows in shielding walls. Special types of glass have been developed to overcome this disadvantage. They contain cerium oxide, and do not discolour on irradiation [28]. High-purity quartz is also resistant to the formation of visible colour on irradiation although absorption bands can develop in the ultraviolet.

The colouration of glass, and the luminescence which appears when the irradiated glass is illuminated with light, have been used as a basis for dosimetry (p. 53). Similar phenomena find applications in geology and related fields for such purposes as the dating of meteorites, minerals and ancient pottery [29]. In the thermoluminescent dating of pottery, for example, the material is first crushed and separated into fractions. The fraction containing colourless minerals such as quartz and felspars is then heated and the luminescence emitted on decay of the optical centres is measured to provide a basis for estimating the age.

Ice Frozen alkaline aqueous glasses develop an intense blue colour on irradiation, the position of the absorption maximum depending on temperature. In 10M NaOH at 77°K, the peak is at 590 nm. The position of the absorption maximum, the electron spin resonance spectrum of the samples and the effect of chemicals which are believed to react with electrons, all show that the colour is associated with trapped electrons. From the size of the ESR signal, the yield of trapped electrons in 10M NaOH at 77°K is close to $G = 1.9$, from which the extinction coefficient at the absorption maximum is 2.0×10^4 M^{-1} cm^{-1}.

The nature of the electron trap has been much debated [30]. Illumination of the coloured ice with 700 nm light causes the absorption maximum of the trapped electron to shift to shorter wavelengths. This shows that there are at least two types of trapped electrons requiring different amounts of energy to enable them to escape from their traps. Electron spin resonance studies show that the traps are associated with the hydrogen atoms of several water molecules, so the electrons may be visualized as being located within cages of oriented water molecules (cf. pp. 147–148), perhaps close to the positive hole. The main function of the alkali may be to enable a glass to form. In the irradiation of crystalline hydrates with the same composition as some of the glasses (for example, NaOH, 3.5 H_2O) different colours are produced in much smaller yield. It may be that in this case the principal trap is a negative ion vacancy, analogous to the case with alkali halide crystals [31].

In the case of the glassy alkaline ice, the positive holes would be expected by inference from results on water irradiated in the vapour and liquid to be the species O^- formed by:

$$H_2O^+ + H_2O \longrightarrow H_3O^+ + OH \tag{5.1}$$

$$OH + OH^- \longrightarrow O^- + H_2O \tag{5.2}$$

Species attributable to O^- have been detected in irradiated alkaline ice by electron spin resonance.

Colour centres are formed in very low yield in crystalline ice at 77°K, $G \sim 10^{-3}$ or less, probably because there are few configurations capable of trapping electrons. In acid glasses at 77°K (for example, several molar $HClO_4$, H_2SO_4 or H_3PO_4), no electrons are seen but hydrogen atoms have been detected by electron spin resonance [32] with G about 1–2. Presumably the hydrogen ions have acted as powerful 'chemical traps', converting electrons into hydrogen atoms according to the reaction:

$$e^- + H_3O^+ \longrightarrow H + H_2O \tag{5.3}$$

analogous to the reactions which are believed to occur in the vapour (p. 124) and liquid (p. 141) phases.

Chemical decomposition

Inorganic solids which contain labile bonds decompose on irradiation, and the mechanism must contain features which are typical of those found in the radiation chemistry of labile substances as well as those typically responsible for radiation effects in simple inorganic solids. A good example is provided by alkali nitrates, which decompose to nitrite and oxygen on irradiation [33]. Excitation and ionization are the dominant primary processes. There is a back reaction owing to the cage effect.

Most inorganic solids however do not decompose. Sulphates, oxides and silicates provide many examples. Solids such as pure metals or graphite which consist of a single element can obviously not decompose in the normal sense. Alkali halides undergo a type of decomposition in that when irradiated crystals are dissolved in water the pH increases, polymerization of vinyl monomers can be initiated and ferrous ions can be oxidized. However, these effects are best understood not in terms of decomposition but in terms of the reactions of the electrons and positive holes trapped at vacancies in the crystal.

The subject of the chemical decomposition of inorganic solids has been reviewed elsewhere [34].

Surface reactions

Radiation can produce characteristic effects on heterogeneous solid-gas or solid-liquid systems. These may originate in radiation-produced changes in the solid or, when present, in the surface film, as well as in radiolytic changes in the gas or liquid if present during irradiation. Examples of practical importance are to be found in the effect of radiation on catalysis [35, 36] (a field of industrial interest where the most important results would very often be kept secret) on metallic corrosion [37] and on the reaction between graphite and carbon dioxide (p. 126) or oxygen in nuclear reactors.

PROBLEMS

1. Using information from Chapter 2, calculate the least energy which fast electrons must have to be capable of producing changes in (*a*) aluminium (*b*) lead. Assume an atom must receive 25 eV to become displaced.

[Ans: (*a*) 0.25 MeV (*b*) 1.13 MeV]

2. In the experiment whose result is given in Fig. 5.2, what percentage of the electron energy absorbed in the sample had been stored in a form which could be released by heating to about 50°K? (assume the stopping power of 1.2 MeV electrons in copper to be 1.2 keV μm^{-1}: density of copper = 8.96 g cm^{-3}).

[Ans: 5.4×10^{-5}]

3. In the experiment whose result is given in Fig. 5.3, how much energy would be liberated on warming the sample of copper which had been irradiated to 10^{16} deuterons per cm^2 to a temperature of around 50°K assuming 5 eV are liberated for each defect responsible for the resistivity change?

[Ans: 0.16 calories per gram]

4. It has been reported that at 30°C solid potassium nitrate is decomposed by γ-rays with $G = 1.5$ while at a similar temperature it is decomposed by 3.4 MeV α-particles with $G = 2.2$. At 150°C γ-rays decompose potassium nitrate with $G = 3.0$ [38]. Suggest reasons for the difference between the two types of radiation at the lower temperatures.

[Ans: see the discussion in Reference 38]

5. If the irradiation of moist air produces nitric acid with $G = 1$, calculate the corrosion rate in mils per year of a 5 cm × 5 cm sheet of nickel (density 8.9 g cm^{-3}) attacked by the nitric acid from 1 litre of moist air of density 1.2 gram per litre irradiated with a dose-rate of 10^6 rads per hour (1 mil = 0.002 54 cm). Assume four molecules of nitric acid are required to convert one atom of nickel to the nitrate and that attack is uniform on both sides of the sheet.

[Ans: 0.14]

REFERENCES

1. M. W. Thompson. *Defects and Radiation Damage in Metals.* Cambridge University Press, Cambridge, 1969.
2. E. W. Müller. *Advan. Electronics Electron Phys.*, **13**, 83–179 (1960), 'Field ionization and field ion microscopy'.
3. S. F. Kurath. *Am. Mineralogist*, **42**, 91–9 (1957), 'Storage of energy in metamict minerals'.
4. C. J. Meechan and A. Sosin. *Phys. Rev.*, **113**, 422–30 (1959), 'Stored energy release in copper following electron irradiation below 20°K'.
5. W. K. Woods, L. P. Bupp and J. F. Fletcher. *U.N. Intern. Conf. Peaceful Uses Atomic Energy 1st Geneva*, 1955, **7**, 455–71, 'Irradiation damage to artificial graphite'.
6. H. G. Cooper, J. S. Koehler and J. W. Marx. *Phys. Rev.*, **97**, 599–607 (1955), 'Irradiation effects in Cu, Ag, and Au near 10°K'.
7. D. S. Billington and J. H. Crawford, Jr. *Radiation Damage in Solids*. Princeton University Press, Princeton, N.J., 1961, p. 314.
8. J. F. Fowler. *Proc. Roy. Soc. (London)*, **A236**, 464–80 (1956), 'X-ray induced conductivity in insulating materials'.
9. S. H. Paine and J. H. Kittel. *U.N. Intern. Conf. Peaceful Uses Atomic Energy 1st Geneva*, 1955, **7**, 445–54, 'Irradiation effects in uranium and its alloys'.
10. J. H. Kittel and S. H. Paine. *U.N. Intern. Conf. Peaceful Uses Atomic Energy 2nd Geneva*, 1958, **5**, 500–9, 'Effect of irradiation on fuel materials'.
11. C. Cawthorne and E. J. Fulton. *Nature*, **216**, 575–6 (1967), 'Voids in irradiated stainless steel'.
12. P. G. Shewmon. *Science*, **173**, 987–91 (1971), 'Radiation-induced swelling of stainless steel'.
13. R. S. Barnes, A. T. Churchman, G. C. Curtis, V. W. Eldred, J. A. Enderby, A. J. E. Foreman, O. S. Plail, S. F. Pugh, G. N. Walton and L. M. Wyatt. *U.N. Intern. Conf. Peaceful Uses Atomic Energy 2nd Geneva*, 1958, **5**, 543–65, 'Swelling and inert gas diffusion in irradiated uranium'.
14. J. A. L. Robertson. *Irradiation Effects in Nuclear Fuels.* Gordon and Breach, New York, N.Y., 1969.
15. J. A. Brinkman, C. E. Dixon and C. J. Meechan. *Acta Met.*, **2**, 38–48 (1954), 'Interstitial and vacancy migration in Cu_3Au and copper'.
16. S. Siegel. *Phys. Rev.*, **75**, 1823–4 (1949), 'Effect of neutron bombardment on order in the alloy Cu_3Au'.
17. F. Fleeman and G. J. Dienes. *J. App. Phys.*, **26**, 652–4 (1955), 'Effect of reactor irradiation on the white-to-grey tin transformation'.
18. M. C. Wittels and F. A. Sherrill. *Phys. Rev. Letters*, **3**, 176–7 (1959), 'Fission fragment damage in zirconia'.

19. M. J. Makin. *Progr. Nuclear Energy, Ser V*, **2**, 500–30 (1959), 'Effects of radiation on mechanical properties of solids'.
20. J. R. Weir Jr. *Science*, **156**, 1689–95 (1967), 'Radiation damage at high temperatures'.
21. J. M. Schulman and W. D. Compton. *Color Centers in Solids*. Pergamon Press, New York, N.Y., 1962.
22. F. Seitz. *Rev. Modern Phys.*, **18**, 384–408 (1946) 'Color centers in alkali halide crystals'.
23. F. Seitz. *Rev. Modern Phys.*, **26**, 7–94 (1954), 'Color centers in alkali halide crystals. II'.
24. H. Dorendorf and H. Pick. *Z. Physik*, **128**, 166–71 (1950), 'Verfärbung von Alkalihalogenidkristallen durch energiereiche Strahlung'.
25. J. J. Markham. *Phys. Rev.*, **88**, 500–9 (1952), 'Speculation on the formation of *F*-centers during irradiation'.
26. For example, J. H. O. Varley. *Nature*, **174**, 886–7 (1954), 'A mechanism for the displacement of ions in an ionic lattice'.
27. D. A. Wiegand and R. Smoluchowski, in *Actions Chimiques et Biologiques des Radiations*, septième série, ed. M. Haïssinsky. Masson, Paris, 1964, pp. 165–256, 'The production of defects in crystalline alkali halides by ionizing radiation'.
28. G. S. Monk. *Nucleonics*, **10** (11) 52–5 (1952), 'Coloration of optical glass by high-energy radiation'.
29. D. J. McDougall, ed. *Thermoluminescence of Geological Materials*. Academic Press, New York, N.Y., 1968.
30. For example, B. G. Ershov and A. K. Pikaev. *Radiat. Res. Rev.*, **2**, 1–101 (1969), 'Stabilized free radicals in the radiation chemistry of frozen aqueous solutions'.
31. A. K. Pikaev, B. G. Ershov and S. A. Puntezhis. *Radiat. Effects*, **5**, 265–8 (1970), 'On the mechanism of low temperature radiolysis of crystalline ice'.
32. R. Livingston, H. Zeldes and E. H. Taylor. *Discussions Faraday Soc.*, **19**, 166–73 (1955), 'Paramagnetic resonance studies of atomic hydrogen produced by ionizing radiation'.
33. For example, J. Cunningham and H. G. Heal. *Trans. Faraday Soc.*, **54**, 1355–69 (1958), 'The decomposition of solid nitrates by X-rays'.
34. E. R. Johnson. *The Radiation-Induced Decomposition of Inorganic Molecular Ions*. Gordon and Breach, New York, N.Y., 1970.
35. R. Coekelbergs, A. Crucq and A. Frennet. *Advan. Catalysis*, **13**, 55–136 (1962), 'Radiation catalysis'.
36. E. H. Taylor. *Advan. Catalysis*, **18**, 111–258 (1968), 'The effects of ionizing radiation on solid catalysts'.
37. J. J. Stobbs and A. J. Swallow. *Met. Rev.*, **7**, 95–131 (1962), 'Effects of radiation on metallic corrosion'.
38. C. J. Hochanadel. *Radiat. Res.*, **16**, 286–302 (1962), Evidence for 'thermal spikes' in the alpha-particle radiolysis of nitrate crystals.

Gases

The responses of molecules to radiation can be understood at a more fundamental level in the gas phase than in the condensed phases for several reasons. First, the energy levels of the system are not affected so much by interactions between molecules, so the primary activations are closer to those which may be calculated for isolated molecules. Second, the yields of ions can be measured much more readily in the gas phase than in the condensed phases. Third, the reactions of ions will be much closer to those studied by techniques like mass spectroscopy. Fourth, the primary excitations and ionizations are widely separated in the gas phase except at high pressures, and spur or track effects are much less important than in the condensed phases. Also it is possible to try to elucidate the role of ions by applying electric fields to modify the effect of the radiation. However, this technique introduces a number of extra features so that it is less informative about the primary processes than hoped, although the features are interesting in themselves. A complication which is present in the gas phase but not in the condensed phases is that the vessel wall may play a part in the radiolysis.

As well as the interest for fundamental studies, the effect of radiation on several gases is of practical importance. For example, air produces ozone and oxides of nitrogen, products which can be dangerous to health and cause corrosion in the neighbourhood of powerful sources of radiation. The effect of radiation on carbon dioxide has to be taken into account in connection with the use of the gas as a coolant in certain types of nuclear reactor. The irradiation of nitrogen-oxygen mixtures, of ethylene and of several other gases has been studied in the hope of devising useful industrial processes. This chapter deals with the irradiation of some individual gases in the light of the general discussion given in earlier chapters, and especially Chapter 4. Some technological aspects are briefly discussed.

Hydrogen

A great deal is known from theoretical and experimental spectroscopy about the electronic states of the hydrogen molecule and its ionized form H_2^+. From this work we could calculate that the primary activation of hydrogen would consist of excitation to defined singlet excited states, some of which would then give rise to hydrogen atoms within the time of a vibration. Some of the atoms would be kinetically and/or electronically excited (hot atoms). H_2^+ would be formed in its ground and excited states and H^+ would appear. To a first approximation the yield of neutral excited hydrogen molecules would be about the same as the yield of ion-pairs which, from direct measurement of W (Table 4.8, p. 78) is $G = 2.75$. The subsequent reactions of the various intermediates in irradiated systems cannot be specified *a priori*, although photochemical, thermodynamic and mass spectroscopic data provide valuable clues. One important reaction is the ion-molecule reaction:

$$H_2^+ + H_2 \longrightarrow H_3^+ + H \qquad (6.1)$$

which was observed long ago in mass spectrometers operated at high pressure.

Now pure hydrogen can hardly undergo permanent chemical change on irradiation. Yet appreciable reaction has been demonstrated when distinguishable forms of the hydrogen molecule have been irradiated. The first work was on the conversion of para- to ortho-hydrogen.† When mixtures of para- and ortho-hydrogen containing more than the equilibrium amount of para- were exposed to irradiation, it was found that para-ortho- conversion occurred with yields as high as about 10^3 molecules converted per ion pair [1]. The high yields show clearly that a chain reaction is taking place. In 1936, after quite detailed analysis, the observations were shown to be consistent with propagation of the chain by atomic hydrogen [2].

$$H + H_2(p) \longrightarrow H_2(o) + H \qquad (6.2)$$

† In the para- state of hydrogen the nuclear spins are anti-parallel; in the ortho- state they are parallel. Para- is the stable form at low temperatures. At ordinary temperatures an equilibrium is attained consisting of 25 per cent para- and 75 per cent ortho-, but the rate of attainment of equilibrium is negligible in the absence of catalysts.

The initiating atoms would be formed by dissociation of neutral excited hydrogen molecules as well as by Reaction 6.1 and neutralization of H_3^+ by electrons. Termination would occur when two H atoms meet at a third body or, especially, at the wall.

Radiation can also induce a chain reaction when mixtures of H_2 and D_2 are irradiated. Addition of small amounts of xenon produces a remarkable decrease in the yield of the reaction [3] (Fig. 6.1). Krypton acts rather

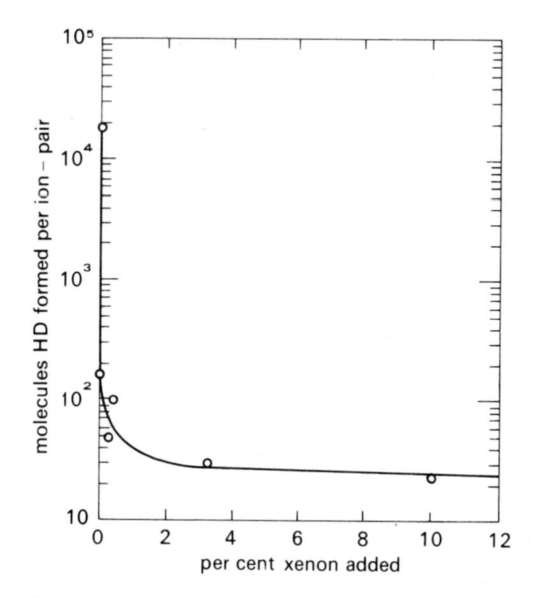

Fig. 6.1 Effect of xenon on the α-particle-induced reaction between H_2 and D_2.

similarly whilst helium, neon and argon produce little effect. These observations can be explained only if krypton and xenon inhibit the propagation steps. Krypton and xenon have no reactivity towards free radicals, but have low ionization potentials and can readily enter into ion-molecule reactions. Hence it seems that an ionic chain reaction is much more important in the radiolysis of hydrogen than a free-radical one as in Equation 6.2. The exact mechanism of the ionic chain reaction is open to question but initiation could occur by Reaction 6.1, occurring with D_2 as well as H_2. This could then be followed by proton transfer reactions such as:

$$H_3^+ + D_2 \longrightarrow H_2 + HD_2^+ \qquad (6.3)$$

$$HD_2^+ + H_2 \longrightarrow HD + DH_2^+ \qquad (6.4)$$

Xenon may inhibit the chain by accepting a proton (or deuteron) from one of the isotopic forms of H_3^+, for example:

$$H_3^+ + Xe \longrightarrow H_2 + HXe^+ \tag{6.5}$$

Neutralization of H_3^+ (or HD_2^+, DH_2^+ or D_3^+) could terminate the chain. When the ionic chain reaction is fully inhibited by xenon or krypton there is still a residual chain reaction, and this no doubt occurs by a hydrogen atom mechanism. The para-ortho- conversion studied earlier must also occur predominantly by the ionic mechanism, unless the ionic chain is inhibited.

When oxygen, even in small proportion, is present in irradiated hydrogen, water is formed with $M/N = 3.6$ at room temperature [4]. The small yield shows that there is no chain reaction. Little is known about the mechanism, but hydrogen peroxide is almost certain to be an intermediate because of the reaction:

$$2HO_2 \longrightarrow H_2O_2 + O_2 \tag{6.6}$$

Hydrogen-chlorine mixtures give rise to hydrogen chloride with yields [5] up to $M/N = 5 \times 10^5$. From parallel studies of the radiation- and photo-induced chain reactions it seems likely that in this system the radiation-induced reaction, like the photochemical reaction, is propagated mainly by the facile free-radical steps:

$$H + Cl_2 \longrightarrow HCl + Cl \tag{6.7}$$

$$Cl + H_2 \longrightarrow HCl + H \tag{6.8}$$

Excitation and ionization of chlorine as well as hydrogen would contribute to the formation of initiating radicals. The threshold energy for the reaction:

$$e^- + Cl_2 \longrightarrow Cl^- + Cl \tag{6.9}$$

is 1.6 eV [6] so that thermalized electrons could not react with chlorine by simple dissociative capture, although they could react through more complex reactions, perhaps involving clusters of ions and molecules. Negative ions could be neutralized by positive ions to form more initiating radicals. Termination may be by the same recombination of H and Cl atoms as occurs in the photochemical case.

An important radiation-induced heterogeneous reaction takes place when tritium gas is allowed to come into contact with organic compounds: the β-rays induce an exchange between the tritium and the hydrogen atoms in the material [7]. This is used as the basis of a method of labelling organic compounds. Typically, a gram of the compound might be left in contact with several curies of tritium gas for some days or weeks, leading to the production of some tens of millicuries of labelled material. The reaction is surprising in view of the high bond strength of H—H. A similar radiation-induced exchange between D_2 and polyethylene has been attributed to a very slow free-radical reaction (p. 217).

Oxygen

The primary activation of oxygen may be represented as follows:

$$O_2 \rightsquigarrow O_2^* \qquad (6.10)$$

$$O_2 \rightsquigarrow 2O \qquad (6.11)$$

$$O_2 \rightsquigarrow O_2^+ + e^- \qquad (6.12)$$

$$O_2 \rightsquigarrow O^+ + O + e^- \qquad (6.13)$$

where as well as O_2^* some of the other species on the right hand side of the equation will be excited. The yield of primary activations to neutral excited states (including the dissociative states corresponding to Equation 6.11) will be approximately equal to the yield of ion pairs, which is $G = 3.1$. O_2^+ accounts for the great majority of positive ions in the mass spectrometer, and this would also be the case in the irradiated system except at very high pressures. Electrons readily react with oxygen, and if the pressure is greater than one tenth of an atmosphere and the dose rate less than 10^{22} eV g^{-1} s^{-1}, all electrons are likely to attach to oxygen in the three-body process:

$$e^- + 2O_2 \longrightarrow O_2^- + O_2 \qquad (6.14)$$

At very low pressures and very high dose-rates some of the electrons may be captured by positive ions [8].

Experimentally it is found that ozone is produced in the irradiation of oxygen in rather variable yield. The variability is caused by difficulties with dosimetry and with the technique for the determination of ozone, and by the instability of ozone which, as well as being affected by impurities, also reverts to oxygen in a radiation-induced chain reaction. Providing care is taken with the techniques, the initial yield can be determined by methods such as irradiating at low temperatures to low doses or measuring the ozone spectroscopically after a short pulse of radiation. Experiments using a 30 ns pulse [9] show that little ozone is formed during the pulse itself but ozone appears afterwards, taking about 30 μs to reach its final concentration, which corresponds to a yield of $G = 13.8$. The rate of formation after the pulse is consistent with ozone formation by reaction of an oxygen atom with an oxygen molecule, with another oxygen molecule acting as a third body to take away the energy liberated in the reaction:

$$O + 2O_2 \longrightarrow O_3 + O_2 \qquad (6.15)$$

If the yield of ozone in these experiments is $G = 13.8$, the yield of oxygen atoms must also be 13.8. It may be assumed that the O_2^+ formed in Reaction 6.12 is neutralized by the O_2^- formed in Reaction 6.14 according to:

$$O_2^+ + O_2^- \longrightarrow 2O + O_2 \qquad (6.16)$$

so that from the yield of ion-pairs, and neglecting O^+, the yield of oxygen atoms from this source is $G = 6.2$. The remaining part of the yield, $G = 7.6$ can be explained if the excitation yield is $G = 3.8$ and every excitation gives two oxygen atoms, either through dissociation of the O_2^* formed in Reaction 6.10 or directly as in Reaction 6.11.

The mechanism of the subsequent radiation-induced chain decomposition of ozone is less clear, but may involve the step:

$$O_2^- + O_3 \longrightarrow O_2 + O_3^- \qquad (6.17)$$

a reaction which would be favoured at low dose-rates, followed by:

$$O_3^- + O_3 \longrightarrow O_2^- + 2O_2 \qquad (6.18)$$

The chain could be terminated by neutralization of O_2^+ by O_3^-. Another possibility is a chain with oxygen atoms as carrier:

$$O + O_3 \longrightarrow 2O + O_2 \qquad (6.19)$$

Water vapour

The excitation of water will lead to the formation of H, OH, H_2 and O, some of which will possess excess kinetic or electronic energy:

$$H_2O \rightsquigarrow H + OH \qquad (6.20)$$

$$H_2O \rightsquigarrow H_2 + O \qquad (6.21)$$

The major positive ion will be H_2O^+

$$H_2O \rightsquigarrow H_2O^+ + e^- \qquad (6.22)$$

OH^+ is seen in the mass spectrum of water with an abundance of 18 per cent so there is a likelihood of its formation under radiolysis conditions too. The yield of excitation has been estimated [10] to be about $G = 4$, while from the value of W for water vapour (Table 4.8) the yield of ion pairs is $G = 3.3$.

H_2O^+ is likely to react at every collision with a water molecule as demonstrated with the mass spectrometer:

$$H_2O^+ + H_2O \longrightarrow H_3O^+ + OH \tag{6.23}$$

Hydronium ions attract various numbers of water molecules round them, depending on temperature and pressure, to form clusters resembling hydrated protons in the liquid phase [11].

Experiments with pure water vapour show that it is rather stable to irradiation when irradiated in rigorously clean reaction vessels. On the other hand appreciable amounts of reaction can be demonstrated in the presence of certain additives [12]. Evidently the irradiation of water vapour leads to decomposition, but there is an efficient back reaction unless the added substances are present. Such reactions have to be taken into account in the design and operation of certain nuclear reactors.

In the presence of small amounts of compounds such as cyclohexane or methanol, water vapour gives hydrogen on irradiation with $G \sim 7.5$.

Table 6.1 *Typical experimental conditions used for irradiation of water vapour* [13]

Type of radiation	cobalt-60 γ-rays
Vessel	glass spherical flask
Vessel volume	5 000 ml
Temperature	116°C
Pressure	500 torr
Dose-rate	10^{14} eV g^{-1} s^{-1}
Irradiation time	15 hours

Typical conditions for experiments leading to this conclusion are listed in Table 6.1. Most of the hydrogen in these experiments is probably formed through reaction of hydrogen atoms with the added substance:

$$H + RH \longrightarrow H_2 + R\cdot \tag{6.24}$$

When the vapour contains added substances such as oxygen or propylene which take up hydrogen atoms without giving hydrogen, the yield of hydrogen can be reduced to as low as $G = 0.5$. This unscavengeable yield of hydrogen has been confirmed in numerous experiments. It may derive from Reaction 6.21, but reactions of hot atoms or other processes could also be responsible, at least in part.

Subtraction of $G = 0.5$ from $G \sim 7.5$ leaves $G \sim 7$. This yield can be accounted for if Reaction 6.20 proceeds with $G = 3.7$ and Reaction 6.22

proceeds with $G = 3.3$, to be followed by Reaction 6.23, and then the hydration of protons, and:

$$e^- + H^+(H_2O)_n \longrightarrow H + (H_2O)_n \qquad (6.25)$$

with all H atoms reacting according to 6.24.

Nitrogen and compounds of nitrogen

Pure nitrogen is necessarily stable to radiation, and not very easy to label. Consequently the radiation chemistry of nitrogen itself has not received extensive study. However, the radiation chemistry of mixtures of nitrogen with other substances, and of compounds of nitrogen, has received a great deal of attention, principally for technological reasons.

Ammonia Mixtures of nitrogen and hydrogen have been found to give ammonia on irradiation. The reverse reaction, the formation of nitrogen and hydrogen by irradiation of ammonia, also takes place. Another product of irradiation of ammonia is hydrazine, a material which finds an important application as a rocket fuel. In the early 1960s considerable attention was devoted to the fission-fragment irradiation of ammonia as a possible method of manufacturing hydrazine on an industrial scale.

The hydrazine has been detected in a yield depending on conditions as high as $G = 4$ [14]. But it does not persist, and on long irradiation at low dose-rates nitrogen and hydrogen in stoichiometric amounts are the only products in the radiolysis of ammonia, the yield at room temperature corresponding to $G = 3.0$ molecules of ammonia decomposed per 100 eV. At higher temperatures the yield increases, reaching a limiting value of $G = 10.0$. Decomposition continues to take place on prolonged irradiation until about 90 per cent of the ammonia is decomposed. The same equilibrium can also be reached from the other side by irradiating nitrogen-hydrogen mixtures.

The mechanism of the radiolysis of ammonia must include the formation of H, NH_2 and NH, all of which have been observed in photochemistry, together with NH_3^+ and possibly NH_2^+, both of which give NH_4^+:

$$NH_2^+ + NH_3 \longrightarrow NH_3^+ + NH_2 \qquad (6.26)$$

$$NH_3^+ + NH_3 \longrightarrow NH_4^+ + NH_2 \qquad (6.27)$$

The hydrazine is formed by dimerization of NH_2 amongst other reactions, and decomposed by radical attack, for example:

$$H + N_2H_4 \longrightarrow H_2 + N_2H_3 \qquad (6.28)$$

$$NH_2 + N_2H_4 \longrightarrow NH_3 + N_2H_3 \qquad (6.29)$$

The ready decomposition of hydrazine by such reactions renders its manufacture by radiation on a commercial scale rather impractical.

Mixtures of nitrogen and oxygen Irradiation of air at atmospheric pressure with single pulses of fast electrons yields ozone with $G = 10.3$ [15]. Little NO_2 is produced. Irradiation of nitrogen-oxygen mixtures under other conditions tends to produce lower yields of ozone and higher yields of NO_2. Other oxides of nitrogen including N_2O_5 and N_2O are formed too. The highest reported yields of NO_2 are $G = 5$–6, produced by irradiation at pressures greater than 100 atmospheres [16]. When water is present, nitric acid can be produced with G up to 2–3. The kinetics of the radiolysis of nitrogen-oxygen mixtures have received extensive investigation and mechanisms involving very large numbers of reactions, some established and some speculative, have been put forward.

Ozone is toxic, and its concentration in a working environment must not be allowed to rise above 0.1 parts per million. Ozone production from air is not serious with most laboratory radiation sources but has to be taken into account with large accelerators [17]. Nitric acid production has been known to cause serious corrosion of accelerators and nuclear reactors: the best safeguards include choice of suitable constructional materials, minimization of the amount of air irradiated and adequate ventilation. The possibility of using radiation for the commercial fixation of nitrogen has received thorough investigation since the original proposal to use fission fragment irradiation for this purpose (p. 4). Very detailed design studies have been made but the results show that in the present state of knowledge the process is not likely to be competitive with conventional methods under prevailing commercial conditions.

Nitrous oxide Nitrous oxide has many features which make it attractive as a gas-phase dosimeter. It is readily available in a pure form, safe, inert, clean, and on irradiation gives a major product, nitrogen, in a yield which is high but not so high as to imply a chain reaction, which would make for irreproducibility. The yield of nitrogen has been measured by several workers. At dose-rates up to 10^{18} eV g^{-1} s^{-1} the yield at ambient temperatures and at pressures 50–600 torr is $G = 10.0 \pm 0.2$ [18]. The yield appears to increase by about one quarter on going to dose-rates as high as 10^{27} eV g^{-1} s^{-1} or on increasing the temperature to 100°C. Many studies of the reaction mechanism have been made. One of the most interesting features is that nitrous oxide captures electrons according to:

$$e^- + N_2O \longrightarrow N_2 + O^- \tag{6.30}$$

a reaction which occurs in the liquid phase as well as in the gas, and has proved invaluable in elucidating reaction mechanisms.

Oxides of carbon

Carbon monoxide In contrast to the gases so far discussed, radiolysis of carbon monoxide yields a solid on irradiation, as well as gaseous products. The solid has an empirical formula which corresponds to a sub-oxide of carbon. The other main product is carbon dioxide, formed with G about 2. The yield for loss of CO is about $G = 8$.

When oxygen is present, the solid no longer appears but a chain reaction can take place, producing carbon dioxide. The mechanism is probably ionic. [19]

Carbon dioxide From photochemical evidence, excited carbon dioxide can give carbon monoxide and oxygen atoms, in various excited states, and possibly carbon atoms. In the mass spectrometer the principal ion seen is CO_2^+, with CO^+, C^+ and O^+ each formed in yields less than 10 per cent of that of CO_2^+. In the irradiated system electrons may be captured by the oxygen which will rapidly appear in small amounts.

Like several of the other gases discussed, carbon dioxide is stable overall to irradiation, although small amounts of carbon monoxide and oxygen can be detected after irradiation at very high dose-rates, or on irradiation of a flowing system. Presumably there is a radiation-induced back reaction comparable to the reaction between carbon monoxide and oxygen referred to above. Decomposition can be promoted by the presence of small amounts of NO_2 (0.5–2 per cent).

Because of its heat transfer and nuclear properties, and its stability to radiation, carbon dioxide makes a good reactor coolant. In a graphite-moderated reactor it then becomes necessary to consider the effect of radiation on the heterogeneous CO_2—C system. In the absence of radiation there is negligible reaction between graphite and carbon dioxide at temperatures below 600°C, but under reactor conditions there is appreciable reaction at all operating temperatures. For example, there are graphite losses as high as 500 kg per year in the Calder-2 power reactor [20]. The overall reaction is:

$$C + CO_2 \longrightarrow 2CO \tag{6.31}$$

In principle the basis of this effect could be either displacement of atoms in graphite by neutrons or changes of a radiation-chemical type in the gas. Displacement effects are certainly capable of increasing the reactivity of graphite with gases, but not at the lower temperature. Moreover γ-irradiation has been shown to be as efficient as reactor irradiation. Hence it is the changes in the gas which are the most important. Expressed in terms of energy absorbed by the gas in the pores of the graphite, the maximum yield of carbon atoms oxidized per 100 eV is $G = 2.35$ [21]. The mechanism seems to be attack on the graphite by short-lived species such as excited carbon dioxide molecules and/or oxygen atoms. Addition of carbon monoxide, hydrogen, water or methane to the gas reduces the rate of reaction.

Methane

The methane molecule contains more atoms than the other molecules so far discussed and undergoes a correspondingly large number of processes on irradiation, yielding numerous products. Determination of the products gives information about the processes. Methane also has the special interest of being the simplest of all organic compounds, as well as being one that can readily be studied in the gas phase.

A great deal of relevant information about methane is available from photochemical and mass spectroscopic studies [22, 23]. Consequences of the excitation of methane have been studied by photolysing it, principally at 147.0 nm (8.4 eV) and 123.6 nm (10 eV). The main reactive species formed in the first instance are $CH_3\cdot$, CH_2, CH and H. At higher energies, photoionization has been seen. The mode of decomposition depends on the energy of the exciting photon, so it is not to be expected that the excited species produced by radiation will decompose in precisely the same way as those produced by light, although there will not be gross differences.

In the absence of added radical scavengers, methyl radicals disappear chiefly by recombination reactions, for example:

$$2CH_3\cdot \xrightarrow{+M} C_2H_6 \tag{6.32}$$

Methylene is considered to insert into methane, giving excited ethane which may decompose into methyl radicals or ethylene and hydrogen, or may lose its excess energy at a third body, depending on the excitation energy of the methylene radical it came from, and on the pressure:

$$CH_2 + CH_4 \longrightarrow C_2H_6^* \tag{6.33}$$

$$C_2H_6^* \longrightarrow 2CH_3\cdot \tag{6.34}$$

$$\longrightarrow C_2H_4 + H_2 \tag{6.35}$$

$$\xrightarrow{+M} C_2H_6 \tag{6.36}$$

The CH radical also inserts into methane:

$$CH + CH_4 \longrightarrow C_2H_5^* \tag{6.37}$$

The excited ethyl radical formed in this reaction must have at least 4 eV more than is needed to break a C—H bond, and at pressures below an atmosphere will give ethylene:

$$C_2H_5^* \longrightarrow C_2H_4 + H \tag{6.38}$$

Hydrogen atoms react rather slowly with methane except at high temperatures ($k \sim 10^{10} \exp(-E/RT)$ $M^{-1}\,s^{-1}$, $E \sim 9-10$ kcal per mole):

$$H + CH_4 \longrightarrow H_2 + CH_3\cdot \tag{6.39}$$

They readily add to ethylene ($k \sim 5 \times 10^8$ $M^{-1}\,s^{-1}$):

$$H + C_2H_4 \longrightarrow C_2H_5\cdot \tag{6.40}$$

and can also react with each other at a third body ($k \sim 10^9 - 10^{10}$ $M^{-2}\,s^{-1}$), and with other radicals.

In the mass spectrum of methane the principal ions are CH_4^+ (47 per cent) CH_3^+ (40 per cent) and CH_2^+ (7.5 per cent), but it is not possible to predict *a priori* to what extent the fragmentation processes would be interfered with by deactivations or ion-molecule reactions under conditions where collisions with molecules can occur. CH_4^+ reacts with methane via a loose complex ($k = 7.4 \times 10^{11}$ $M^{-1}\,s^{-1}$):

$$CH_4^+ + CH_4 \longrightarrow CH_5^+ + CH_3\cdot \tag{6.41}$$

CH_3^+ gives $C_2H_5^+$, via an intermediate complex, $C_2H_7^+$ ($k = 10^{12}$ M^{-1} s^{-1}):

$$CH_3^+ + CH_4 \longrightarrow C_2H_5^+ + H_2 \tag{6.42}$$

The reactions of CH_2^+ are less simple, but it is of course a relatively minor species. All of the ion-molecule reactions of 'primary' ions occur before neutralization can take place.

CH_5^+ and $C_2H_5^+$ do not react with methane to give chemically distinguishable products, although CH_5^+ can transfer a proton which in the irradiation of mixtures of CH_4 and CD_4 leads to isotopic mixing. CH_5^+ and $C_2H_5^+$ disappear by reacting with additives (if present) or radiolysis products, or by becoming neutralized.

In the mid-nineteen twenties it was observed that the pressure did not change very much when methane was irradiated, which led to the conclusion that methane, like several other gases, did not undergo any net change on radiolysis. However, later work using mass spectroscopy, gas chromatography and other analytical techniques, revealed substantial decomposition. The composition of the products alters as the irradiation proceeds since as ethylene builds up it begins to be attacked by H atoms, and higher hydrocarbons, even when present at concentrations as low as 0.01 per cent, are attacked by CH_5^+ (proton transfer followed by decomposition) and $C_2H_5^+$ (hydride transfer) [24]. Typical results, for conversion to greater than 0.2 per cent, are shown in Table 6.2.

Table 6.2 *Products formed by irradiation of methane [25] (1.9 MeV electrons, pressure 731 torr)*

Product	*G*
H_2	5.73
C_2H_4	0.004
C_2H_6	2.20
C_3H_6	0.00
C_3H_8	0.36
n-C_4H_{10}	0.114
i-C_4H_{10}	0.040
C_5—C_6	0.03
Polymer	2.1 molecules of CH_4 converted to polymer

From measurements of *W*, the yield of ion pairs in the radiolysis of methane is close to $G = 3.6$. The ratio of excitation to ionization is in the region 0.63–0.8 [26]. If we take 0.7, the yield of excitations becomes $G = 2.5$.

The yield of CH_4^+ can be estimated from experiments in which, for example, 0.3–5 per cent *i*-C_4D_{10} is present in irradiated methane [24]. Methanium ions formed by Reaction 6.41 react with C_4D_{10} principally according to:

$$CH_5^+ + C_4D_{10} \longrightarrow CH_4 + C_4D_{10}H^+ \tag{6.43}$$

followed by:

$$C_4D_{10}H^+ \longrightarrow C_3D_7^+ + CD_3H \tag{6.44}$$

$$C_3D_7^+ + C_4D_{10} \longrightarrow C_3D_8 + C_4D_9^+ \tag{6.45}$$

The yield of C_3D_8 together with a small allowance for other reactions gives an estimate for the yield of CH_4^+. At a methane pressure of 480 torr the value found was $G = 1.9$.

The yield of CH_3^+ can be estimated from the same experiments. All CH_3^+ would be expected to give ethyl ions by reaction 6.42. Ethyl ions react with, for example, *i*-C_4D_{10} by transfer of a hydride ion:

$$C_2H_5^+ + C_4D_{10} \longrightarrow C_2H_5D + C_4D_9^+ \tag{6.46}$$

If the intermediate $C_2H_7^+$ were to react with C_4D_{10} it would form the same final products. From measured yields of C_2H_5D, the yield of CH_3^+ is $G = 0.9$.

The values 1.9 and 0.9, in conjunction with G(total ions) = 3.6, correspond to a fragmentation pattern of methane at 480 torr of CH_4^+ (53 per cent) CH_3^+ (25 per cent), that is, somewhat less fragmentation of the parent ion than in the mass spectrometer. At higher pressures (up to at least 15 atmospheres) the yield of parent ions increases still further, and the CH_3^+ yield decreases correspondingly [27].

It has been stated that the yield of methyl radicals in irradiated methane is $G = 3.3$ [23]. If we assume that the only sources of methyl radicals are excited ethane and Reaction 6.41, then, if $G(CH_4^+)$ is 1.9, the yield of methyl radicals formed by excitation becomes $G = 1.4$. The yield of hydrogen atoms produced in the primary activation may be estimated from results of irradiating CD_4 (40 torr) containing up to 4.5 per cent H_2S [28]. In this system D atoms from CD_4 will form HD according to:

$$D + H_2S \longrightarrow HD + SH \qquad (6.47)$$

No other reaction is likely to produce HD. From the HD yield, the yield of D atoms from CD_4 is $G = 4$. The value for H atoms from CH_4 would not be very different. By material balance, a total of $G = 1.4 + 0.9 = 2.3$ of these may be associated with the primary production of $CH_3\cdot$ and CH_3^+ respectively, and the rest will come from the fragmentation processes which give CH_2, CH, CH_2^+ and minor species.

The yield of molecular hydrogen can be estimated from results of irradiating mixtures of CH_4 and CD_4 containing iodine to scavenge H atoms [29]. In this system hydrogen is formed with $G = 3.2$, H_2, HD and D_2 being in a ratio of approximately 3:1:2 (the exact ratio depends on the experimental conditions). All of the observed HD may be taken to arise from reactions of methyl ions with methane. This reaction will also produce some H_2 and D_2, the amount of which can be estimated from the amount of HD. The rest of the H_2 and D_2 comes from elimination of molecular hydrogen associated with the primary production of CH_2, CH, CH_2^+ and minor species. The yield of this molecular hydrogen is $G = 1.3$.

The yields of CH_2 and CH have been estimated to be 0.7 and 0.1–0.3, respectively [23]. These and other yields can readily be fitted into equations giving material balances for the primary processes although, for those carbon-containing primary species which are deficient in two or more hydrogen atoms, it is not possible to say every time whether the hydrogen has been eliminated in the molecular or atomic forms. The products formed under particular radiolysis conditions will be the results of processes like those discussed, with neutralization being the ultimate fate of the cations. The exact nature of the neutralization reactions can only be guessed.

Ethylene

Photochemical evidence shows that excited ethylene can give C_2H_2 (acetylene), C_2H_3 (vinyl radicals), H_2 and H. The principal positive ions in the mass spectrum of ethylene are $C_2H_4^+$ (38 per cent), $C_2H_3^+$ (23 per cent) and $C_2H_2^+$ (22 per cent): H_2 and H must be produced at the same time.

Vinyl radicals are likely to add to ethylene:

$$C_2H_3 + C_2H_4 \longrightarrow C_4H_7 \qquad (6.48)$$

So are hydrogen atoms (Reaction 6.40).

The ions are likely to add to ethylene too, forming complex ions in the first instance, for example:

$$C_2H_4^+ + C_2H_4 \longrightarrow [C_4H_8^+] \qquad (6.49)$$

At very low pressures (<0.2 torr) the complex ions will dissociate, for example:

$$[C_4H_8^+] \longrightarrow C_3H_5^+ + CH_3\cdot \qquad (6.50)$$

but at pressures above a few torr they can become stabilized by collision [23], for example:

$$[C_4H_8^+] + C_2H_4 \longrightarrow C_4H_8^+ + C_2H_4 \qquad (6.51)$$

Except at very low pressures, the principal final product of the irradiation of ethylene is a polymer, and hydrogen and numerous saturated and unsaturated hydrocarbons are formed too. The free radicals and the ions could both contribute to the propagation of the polymer chain but, since the polymer is precipitating out of the gas phase while it is growing, and once formed is rather intractable, the conditions are not ideal for kinetic studies and it becomes difficult to hope for a complete account of all the major processes taking place. On balance, however, the polymerization seems to take place predominantly by a free-radical mechanism.

Whatever the mechanism, the polymerization of ethylene by radiation has attracted attention in view of the possibility of using the process in industry. The flow sheet of one pilot plant, using a ^{60}Co source of 10^5 curies and a reaction vessel of 10 litres is shown in Fig. 6.2. The pressure was 440 atmospheres and the temperature 150°C. This system produced a few kilograms of polymer in each run [30].

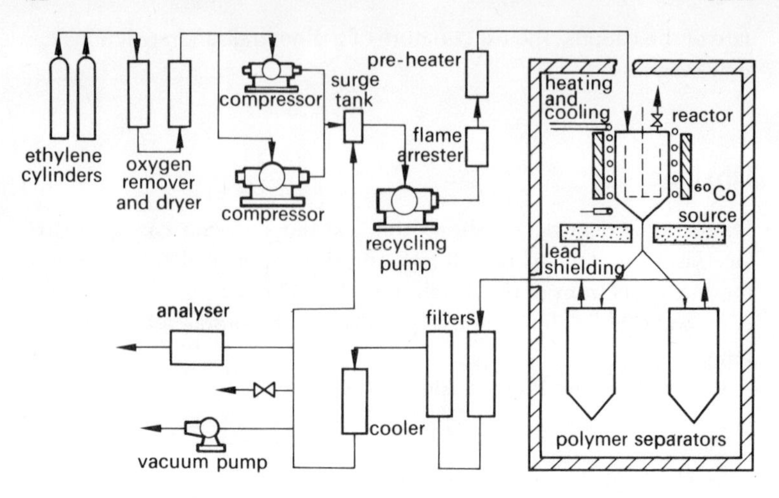

Fig. 6.2 Flow sheet for polymerization of ethylene on a pilot-plant scale.

Another reaction of industrial interest is the irradiation of mixtures of ethylene and hydrogen bromide to form ethyl bromide. This was among the first radiation processes to be carried out commercially. The reaction proceeds by a chain mechanism in which the propagating steps are:

$$Br + C_2H_4 \longrightarrow \dot{C}H_2CH_2Br \quad (6.52)$$

$$\dot{C}H_2CH_2Br + HBr \longrightarrow CH_3CH_2Br + Br \quad (6.53)$$

The commercial process uses a two-phase system in which the reactant gases are passed into liquid ethyl bromide, the new product overflowing out of the reaction vessel [31].

Other gases

Besides those already discussed, numerous other hydrocarbon [22, 23] and other gases [23, 32] have been irradiated. Information from related fields like photochemistry and chemical kinetics has been correlated with the radiation-chemical data, and aspects of the mechanisms have been worked out. Perhaps the greatest value of such studies has been the contribution made to the chemistry of short-lived species. In addition, the radiolysis of numerous gaseous systems has been investigated from a technological point of view, although the results are not always made available to the public.

PROBLEMS

1. If the ionic yield of HD in the irradiation of an equimolar mixture of H_2 and D_2 is 3.0×10^5, what would be the ionic yield for a mixture containing H_2 and D_2 in the molar ratio 3:1 if the chain length is the same?

[Ans: 1.5×10^5]

2. (*a*) Calculate the yield to be expected for the formation of HD in the irradiation of H_2O vapour containing a very small percentage of D_2 if the mechanism comprises reactions expected from the discussion on pp. 122–124 together with:

$$OH + D_2 \longrightarrow HDO + D$$

$$H + D_2 \longrightarrow HD + D$$

$$2D \xrightarrow{M} D_2$$

(neglect Reaction 6.21 and the formation of OH^+ and assume the yields of Reactions 6.20 and 6.22 are 4.0 and 3.3, respectively). (*b*) Seemingly reliable experiments [12] give yields of $G(HD) = 11$. Try to account for the discrepancy.

[Ans: (*a*) $G = 7.3$ (*b*) some possibilities are mentioned on p. 283 of Reference 12]

3. (*a*) If ozone is produced in irradiated air with $G = 10$, how many cm^3 at standard temperature and pressure would be produced in one minute by an accelerator delivering 10 *W* to air? (*b*) If the ozone were dissipated homogeneously throughout the room and the concentration had to be kept to 0.1 parts per million by ventilation, how many air changes per hour would be needed while the accelerator were running continuously if the volume of the room were 10^3 m^3.

[Ans: (*a*) 14 (*b*) 8]

4. Calculate the percentage of the weight of the graphite (bulk density 1.72 g cm^{-3}) which would be expected to be lost per year in a reactor cooled with CO_2 (pressure 7 atmospheres) and moderated with graphite, at a location where the energy absorption in the graphite is 30 mW g^{-1} and the temperature 350°C if the open pore volume of the graphite is 20 cm^3 per 100 cm^3.

[Ans: 0.20]

5. If hydrogen atoms are produced in a mixture of methane (725 torr) and ethylene (35 torr) what would be the approximate ratio of abstractions from methane to additions to ethylene at (*a*) 500°K, (*b*) 700°K, assuming the data on p. 128 with an activation energy of 9 kcal per mole.

(c) What would be the ratios at the same two temperatures if the pre-exponential factor in the expression for reaction of hydrogen atoms with methane were 1.5×10^{11} and the activation energy were 14 kcal per mole.

[Ans: (a) 5×10^{-2} (b) 0.7 (c) 5×10^{-3}, 0.3]

REFERENCES

1. P. C. Capron. *Ann. Soc. Sci. Bruxelles,* **55**, 222–36 (1935), 'La conversion de l'ortho-para hydrogène sous l'action des particules α'.
2. H. Eyring, J. O. Hirschfelder and H. S. Taylor. *J. Chem. Phys.,* **4**, 479–91 (1936), 'The theoretical treatment of chemical reactions produced by ionization processes Part I. The ortho-para- hydrogen conversion by alpha-particles'.
3. S. O. Thompson and O. A. Schaeffer. *J. Am. Chem. Soc.,* **80**, 553–8 (1958), 'The role of ions in the radiation induced exchange of hydrogen and deuterium'.
4. S. C. Lind. *Radiation Chemistry of Gases.* Reinhold, New York, N.Y., 1961.
5. F. Porter, D. C. Bardwell and S. C. Lind. *J. Am. Chem. Soc.,* **48**, 2603–18 (1926), 'The photo- and radiochemical interaction of hydrogen and chlorine'.
6. D. C. Frost and C. A. McDowell. *Can. J. Chem.,* **38**, 407–20 (1960), 'The ionization and dissociation of some halogen molecules by electron impact'.
7. K. E. Wilzbach. *J. Am. Chem. Soc.,* **79**, 1013 (1957), 'Tritium-labeling by exposure of organic compounds to tritium gas'.
8. K. Fueki and J. L. Magee. *Discussions Faraday Soc.,* **36**, 19–34 (1963), 'Reactions in tracks of high energy particles. Radiolysis of oxygen'.
9. J. A. Ghormley, C. J. Hochanadel and J. W. Boyle. *J. Chem. Phys.,* **50**, 419–23 (1969), 'Yield of ozone in the pulse radiolysis of gaseous oxygen at very high dose rates. Use of this system as a dosimeter'.
10. I. Santar and J. Bednář. *Collection Czech. Chem. Commun.,* **32**, 953–67 (1967), 'Theory of radiation chemical yield. I. Radiolysis of water vapour'.
11. P. Kebarle, S. K. Searles, A. Zolla, J. Scarborough and M. Arshadi. *J. Am. Chem. Soc.,* **89**, 6393–9 (1967), 'The solvation of the hydrogen ion by water molecules in the gas phase. Heats and entropies of solvation of individual reactions: $H^+(H_2O)_{n-1} + H_2O \longrightarrow H^+(H_2O)_n$'.
12. R. S. Dixon. *Radiat. Res. Rev.,* **2**, 237–96 (1970), 'The dissociation of water vapour by photolytic, radiolytic and electron impact methods'.
13. J. H. Baxendale and G. P. Gilbert. *Trans. Faraday Soc.,* **36**, 186–92 (1963), 'The γ-radiolysis of water vapour'.
14. F. T. Jones, T. J. Sworski and J. M. Williams. *Trans. Faraday Soc.,* **63**, 2426–34 (1967), 'Radiation chemistry of gaseous ammonia. Part 2. Hydrazine formation'.
15. C. Willis, A. W. Boyd and M. J. Young. *Can. J. Chem.,* **48**, 1515–25 (1970), 'Radiolysis of air and nitrogen-oxygen mixtures with intense electron pulses: determination of a mechanism by comparison of measured and computed yields'.
16. M. T. Dmitriev and S. Ya Pshezhetskii. *Russ. J. Phys. Chem.,* **34**, 418–22 (1960), 'The radiation oxidation of nitrogen. V. Kinetics of the oxidation of nitrogen under the influence of γ-radiation and the part played by ion recombination'.
17. L. N. Less and A. J. Swallow. *Nucleonics,* **22** (9), 58–61 (1964), 'Estimating the hazard due to radiolytic products from air'.

18. F. T. Jones and T. J. Sworski. *J. Phys. Chem.*, **70**, 1546–52 (1966), 'Nitrous oxide dosimetry. Effects of temperature, pressure and electric field'.
19. P. G. Clay, G. R. A. Johnson and J. M. Warman. *Discussions Faraday Soc.*, **36**, 46–55 (1963), 'γ-ray-induced oxidation of carbon monoxide: evidence for an ionic chain-reaction'.
20. A. R. Anderson, H. W. Davidson, R. Lind, D. R. Stranks, C. Tyzack and J. Wright. *U.N. Intern. Conf. Peaceful Uses Atomic Energy 2nd Geneva*, 1958, **7**, 335–73, 'Chemical studies of carbon dioxide and graphite under reactor conditions'.
21. R. Lind and J. Wright. *U.N. Intern. Conf. Peaceful Uses Atomic Energy 3rd Geneva*, 1964, **9**, 541–9, 'Factors controlling reaction between graphite and radiolysed carbon dioxide'.
22. P. Ausloos and S. G. Lias, in *Actions Chimiques et Biologiques des Radiations*, onzième série, ed. M. Haïssinsky. Masson, Paris, 1967, pp. 1–83, 'Gas phase radiolysis of hydrocarbons'.
23. G. G. Meisels, in *Fundamental Processes in Radiation Chemistry*, ed. P. Ausloos. Interscience, New York, N.Y., 1968, pp. 347–411, 'Organic gases'.
24. P. Ausloos, S. G. Lias and R. Gorden, Jr. *J. Chem. Phys.*, **39**, 3341–8 (1963), 'Effect of additives on the ionic reaction mechanism in the radiolysis of methane'.
25. L. W. Sieck and R. H. Johnsen. *J. Phys. Chem.*, **67**, 2281–88 (1963), 'Some aspects of the radiation chemistry of methane'.
26. R. L. Platzman. *The Vortex*, **23**, 372–85 (1962), 'Superexcited states of molecules, and the primary action of ionizing radiation'.
27. P. Ausloos, R. Gorden, Jr. and S. G. Lias. *J. Chem. Phys.*, **40**, 1854–60 (1964), 'Effect of pressure in the radiolysis and photolysis of methane'.
28. R. Gorden, Jr. and P. Ausloos. *J. Chem. Phys.*, **46**, 4823–34 (1967), 'Gas–phase photolysis and radiolysis of methane. Formation of hydrogen and ethylene'.
29. P. J. Ausloos and S. G. Lias. *J. Chem. Phys.*, **38**, 2207–14 (1963), 'Radiolysis of methane'.
30. A. Danno, in *Actions Chimiques et Biologiques des Radiations*, treizième série, ed. M. Haïssinsky. Masson, Paris, 1969, pp. 145–224, 'Industrial applications of radiation chemistry'.
31. D. E. Harmer, J. S. Beale, C. T. Pumpelly and B. W. Wilkinson, in *Industrial Uses of Large Radiation Sources*. IAEA, Vienna, 1963, **2**, pp. 205–28, 'The Dow ethyl bromide process: an industrial application of radiation chemistry'.
32. A. R. Anderson, in *Fundamental Processes in Radiation Chemistry*, ed. P. Ausloos. Interscience, New York, N.Y., 1968, pp. 281–345, 'Inorganic gases'.

Water and aqueous solutions

As is already apparent from earlier chapters, the primary excitations and ionizations produced by radiation, and the subsequent reactions of the species formed, can be significantly different in the condensed phases from in the gas phase. Many of the principles of radiation chemistry as applied to the liquid phase have been developed in the course of studies on the radiation chemistry of water and aqueous solutions. Water and aqueous solutions have been studied because of the part they play in chemistry in general and in radiochemistry in particular, because they are readily available and not too difficult to work with and because water is a polar liquid which responds in characteristic ways to radiation. A practical motivation for the studies has been the desire to understand the effect of radiation on biological systems. Also the irradiation of water and several aqueous systems is an important consideration in various aspects of nuclear technology. As a result of all the work done, many of the aspects of the radiation chemistry of water and aqueous solutions (although by no means all) are now reasonably well understood [1].

Experimental facts

The principal experimental facts had already begun to be acquired by the early years of the twentieth century. Liquid water itself, when highly purified and irradiated under conditions where gas cannot escape from solution, resembles water vapour in that it does not decompose significantly under irradiation with low LET radiation such as X-rays [2]. On the other hand water decomposes into hydrogen, hydrogen peroxide, and oxygen on irradiation with high LET radiation like α-particles. Among the other facts, many of which were confirmed by the radiation chemists of the United States atomic bomb project [3], is that decomposition under low LET irradiation is very much enhanced by impurities. Oxygen is one such 'impurity': it causes water to give hydrogen peroxide and some hydrogen on irradiation with low LET radiation. Hydrogen peroxide itself enhances radiolysis by low LET radiation, but hydrogen tends to suppress radiolysis. Correspondingly if water is irradiated under conditions where hydrogen can escape, for example, in contact with a large evacuated space, or while boiling (for example, in a boiling water reactor), then it undergoes decomposition.

Intermediates in the radiolysis of water

The radiolysis of water is to be understood in terms of the physical processes taking place. The consequences of the physical processes cannot be deduced from first principles, but thoughtful experimental work has enabled the present picture to be built up. The basic experimental observation is that for numerous solutions of different compounds in water the solute becomes affected by the irradiation, the number of solute molecules affected being proportional to the dose delivered but essentially independent of solute concentration. For example, in the X-irradiation of aqueous solutions of ferrous sulphate, ferric ions are produced, and the total amount of ferric ion produced by a given dose is independent of ferrous concentration [4]. This shows that the solute is not being affected directly by the radiation, but is being affected indirectly by some entity or entities produced from the water. This is of course quite plausible, since the greater part of the interaction of the radiation with the system is with the component present in excess, that is, with the water itself.

From very large numbers of experiments, the action of radiation on aqueous solutions consists of reduction or oxidation. This could be understood if the radiation acts on the water to form H atoms (reducing) and OH radicals (oxidizing) [5]. These could be produced by dissociation of excited water molecules or via ionization. Now if the effect of radiation on water is simply to produce H atoms and OH radicals which may then react with solute it might be expected that quite low concentrations of suitable solutes would be sufficient to react with all the atoms and radicals, so that H atoms would then not be able to react with each other to give molecular hydrogen, and OH radicals could not give hydrogen peroxide. Several aqueous solutions, however, including dilute hydrogen peroxide and oxygen-saturated ferrous sulphate, are found to give a certain yield of hydrogen on irradiation whatever the solute concentration. An approximately equivalent amount of hydrogen peroxide also appears to be formed. This can be understood if in addition to decomposition of water into H atoms and OH radicals the radiation also gives some molecular hydrogen and hydrogen peroxide as 'primary' products [6].

Although the reducing species may be represented formally as a hydrogen atom, the reductions can equally be explained if the reducing entity is an electron [7] (the two are related through the acid base equation $e^- + H^+ \rightleftharpoons H$). Some experiments on the X-irradiation of aqueous solutions containing 0.1N acid and low concentrations of methanol and ferric ions were among the first to enable the two to be distinguished [8]. In this work, hydrogen gas and ferrous ions had been detected as products. A small part of the hydrogen found could be assumed to be the 'primary' yield, but most of the hydrogen was believed to derive from the reaction:

$$H + CH_3OH \longrightarrow H_2 + \dot{C}H_2OH \qquad (7.1)$$

Hydroxyl radicals were believed to react by an analogous reaction:

$$OH + CH_3OH \longrightarrow H_2O + \dot{C}H_2OH \qquad (7.2)$$

Both H atoms and $\dot{C}H_2OH$ radicals could reduce ferric ions to ferrous.

$$H + Fe^{3+} \longrightarrow H^+ + Fe^{2+} \qquad (7.3)$$

$$\dot{C}H_2OH + Fe^{3+} \longrightarrow HCHO + H^+ + Fe^{2+} \qquad (7.4)$$

On this mechanism the number of hydrogen molecules detected per 100 eV absorbed in the solution, $G(H_2)$, should be given by the expression:

$$G(H_2) = G(H)\frac{k_1[CH_3OH]}{k_1[CH_3OH] + k_3[Fe^{3+}]} + G(M) \qquad (7.5)$$

where $G(H)$ is the number of hydrogen atoms produced from the water per 100 eV absorbed in the system, k_1 and k_3 are the rate constants for Reactions 7.1 and 7.3, respectively, and $G(M)$ is the yield of 'primary' molecular hydrogen. $G(M)$ was found by measuring $G(H_2)$ at a high value of $[Fe^{3+}]/[CH_3OH]$. By rearranging Equation 7.5 it can be seen that it should be possible to obtain a straight line by plotting a graph of the reciprocal of $G(H_2) - G(M)$ against the ratio of ferric concentration to methanol concentration. Values of k_1/k_3 and $G(H)$ could be obtained from the graph. The methanol concentration was varied for a ferric concentration of 5×10^{-3}M and it was found that the measured hydrogen yield did indeed give a good straight line, from which the values $k_1/k_3 = 0.45$ and $G(H) = 3.2$ were obtained. However, when the methanol concentration was varied for a ferric concentration of 5×10^{-2}M, $G(H)$ was found to be 1.7, although k_1/k_3 remained the same. Taken at face value this result would imply that the yield of ferrous produced should be smaller for 5×10^{-2}M ferric solutions than for 5×10^{-3}M. Actually it was a little greater. This can be explained if at high ferric concentrations, the ferric is being reduced by electrons rather than directly or indirectly by hydrogen atoms as such.

Additional evidence for different forms of the reducing entity derives from work on the effect of X-rays on dilute solutions of chloracetic acid [9]. In this system the main products found are H_2, Cl^- and H_2O_2. The

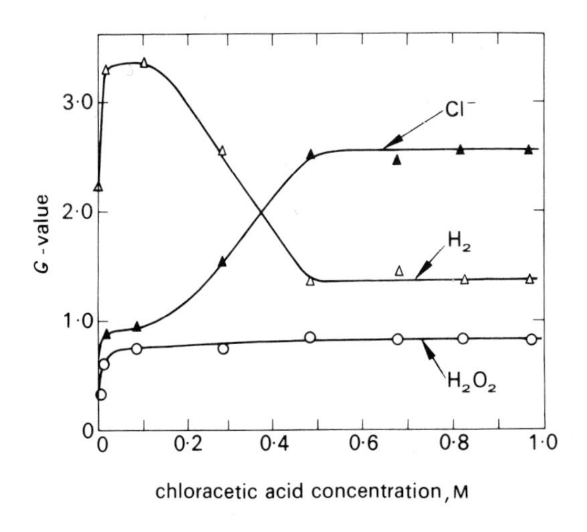

Fig. 7.1 Yields of products in the X-irradiation of aqueous solutions containing chloracetic acid at various concentrations (pH 1).

yields of these products as a function of solute concentration are shown in Fig. 7.1. The hydrogen peroxide yield is practically independent of concentration, consistent with its being a 'primary' product as proposed above. The hydrogen yield is well above the 'primary' yield of molecular hydrogen as deduced from the irradiation of many other solutions. Hydrogen atoms are reactive entities and could easily dehydrogenate chloracetic acid, thus accounting for large yields of hydrogen:

$$H + ClCH_2COOH \longrightarrow H_2 + Cl\dot{C}HCOOH \qquad (7.6)$$

They could also react with chloracetic acid to form HCl and the $\dot{C}H_2COOH$

radical. But why should the chloride yield increase with chloracetic acid concentration as the hydrogen yield goes down? One possibility might have been that Reaction 7.6 does not occur, but at high chloracetic acid concentrations H atoms give HCl, while at lower concentrations the H atoms react with H^+ to give H_2^+ which then forms H_2 by reaction with chloracetic acid. However an alternative, and simpler, explanation would be that the chloride is formed by the action of an electron. In the original paper this was called the 'polaron', and represented as H_2O^-:

$$H_2O^- + ClCH_2COOH \longrightarrow Cl^- + \dot{C}H_2COOH + H_2O \qquad (7.7)$$

The shape of the curves in Fig. 7.1 could be accounted for by competition between Reaction 7.7 and:

$$H_2O^- + H_3O^+ \longrightarrow H + 2H_2O \qquad (7.8)$$

followed by Reaction 7.6.

There is a great deal of other evidence that there are at least two forms of the reducing entity produced by radiation. One form is converted into the other by reaction with acid. However it remains to be decided whether the two forms are indeed the electron and the hydrogen atom rather than some other pair such as the hydrogen atom and H_2^+. Now the rate constants for the reactions of charged species can be shown to depend on ionic strength. For reactant ions of formula A^{Z_a} and B^{Z_b} the dependence may be expressed by the Brønsted-Debye equation:

$$\log_{10}\frac{k}{k_0} = 1.02\, z_a z_b \frac{\mu^{1/2}}{1 + a\mu^{1/2}} \qquad (7.9)$$

where k is the rate constant at ionic strength μ, k_0 is the rate constant at infinite dilution of ions and a is a parameter depending on the distance of closest approach between A and B. The value of a is expected to lie between 1 and 3. At low ionic strength the value adopted makes little difference, and for general use it may be assumed that $a = 2$. Evidence for the charge on the primary reducing entity has been obtained by studying the effect of ionic strength on the relative rate constants for the reactions of the primary reducing species with hydrogen peroxide (k_1), oxygen (k_2), hydrogen ions (k_3) and nitrite (k_4) [10]. It has been found that k_2/k_1 is unaffected by ionic strength, while k_3/k_1 decreases with increasing ionic strength, and k_4/k_1 increases. This shows that the reducing species is negatively charged. From the slopes of the ionic strength curves, the charge is found to be unity. Hence the primary reducing species is likely to be an electron. A similar conclusion has been drawn from a study of the effect of ionic strength on the ratio of the rate constants for reduction of silver ions and acrylamide [11].

Final proof of the nature of the reducing species in irradiated water is obtained from pulse radiolysis: at the conclusion of a microsecond pulse of radiation, a strong absorption is seen, with a maximum at 720 nm (Fig. 4.11 p. 92). The resemblance between this spectrum and that of alkali metals dissolved in liquid ammonia shows it to be due to a hydrated electron.

The evidence so far shows conclusively that hydrated electrons are produced in the radiolysis of water, but it does not follow that no other reducing species is formed at the same time. Work on the radiolysis of aqueous solutions of isopropanol indicates that some hydrogen atoms are produced in addition to the hydrated electrons [12]. When acid solutions of isopropanol are irradiated, hydrogen is produced, as would be expected if the hydrated electrons react with hydrogen ions (cf. Equation 7.8):

$$e^-_{aq} + H_3O^+ \longrightarrow H + H_2O \qquad (7.10)$$

followed by:

$$H + (CH_3)_2CHOH \longrightarrow H_2 + (CH_3)_2\dot{C}OH \qquad (7.11)$$

Above pH ~ 7 the hydrogen yield is markedly less, mainly because the hydrated electron is reacting with one of the irradiation products, acetone. But the yield is still found to be about twice that expected on the basis of the yield of non-scavengeable 'primary' molecular hydrogen as found with, for example, inorganic solutes. Furthermore adding extra acetone before irradiation does not diminish the yield. This can be understood if we always have a certain primary formation of hydrogen atoms as well as hydrated electrons. In the present system these would give molecular hydrogen by Reaction 7.11.

If the principal reducing species in the radiolysis of water is an electron, is it not possible that the main oxidizing species is H_2O^+ (that is, $OH + H^+$) rather than OH? Evidence for the charge on the oxidizing species has been sought by irradiating neutral solutions containing ethanol and bromide ion, both of which react with the oxidizing species [13]. It is found that inert salts do not affect the competition between the two for the oxidizing species which shows that the latter is uncharged.

The available evidence is now consistent with the view that the chemical effects produced on irradiation of water and dilute aqueous solutions are a consequence of the production of the radicals e^-_{aq}, OH and H and the molecules H_2 and H_2O_2 (together with hydrogen ions to balance the equation):

$$H_2O \leadsto e^-_{aq}, OH, H, H_2, H_2O_2, H_3O^+ \qquad (7.12)$$

With X-, γ-rays, fast electrons and low LET radiations generally, the radical intermediates predominate, while with α-particles and other high LET radiations the molecular products are more important.

The hydrated electrons presumably derive from hydration of the electrons formed by ionization. Hydroxyl radicals and hydrogen ions could be formed principally by the reaction:

$$H_2O^+ + H_2O \longrightarrow H_3O^+ + OH \tag{7.13}$$

as in the vapour phase (p. 123). The hydrogen atoms may arise in part from dissociation of excited water molecules and in part from Reaction 7.10 occurring in regions of the particle track where hydrated electrons and hydrogen ions are present together. The large yield of the molecular intermediates in high LET irradiation may arise from the fact that the tracks are dense (Fig. 2.8, p. 31) so that the various species would often interact with each other before they had a chance to diffuse away to react with the solute:

$$e^-_{aq} + e^-_{aq} \xrightarrow{+2H_2O} H_2 + 2OH^- \tag{7.14}$$

$$e^-_{aq} + H_3O^+ \longrightarrow H + H_2O \tag{7.10}$$

$$e^-_{aq} + H \xrightarrow{+H_2O} H_2 + OH^- \tag{7.15}$$

$$2H \longrightarrow H_2 \tag{7.16}$$

$$2OH \longrightarrow H_2O_2 \tag{7.17}$$

$$e^-_{aq} + OH \longrightarrow OH^- \tag{7.18}$$

$$H + OH \longrightarrow H_2O \tag{7.19}$$

These equations imply the transient formation of OH^- as well as the other species, and this species has been detected experimentally. The molecular products formed in low LET irradiation may arise in part from Reactions 7.14–7.17 occurring in spurs, blobs and so on, but several items of evidence, including attempts to account quantitatively for experimental results by diffusion kinetics (pp. 82–83) show that the whole of these products cannot be accounted for in this way, so that a part at least (perhaps 30 per cent in the case of the molecular hydrogen) must originate in other processes.

Yields of intermediates

If the radiolysis of water and dilute aqueous solutions is really to be understood in the terms discussed, then it should be possible to use the measured amount of chemical change to deduce the yields for the 'primary' decomposition of water, Reaction 7.12. The most numerous and meaningful studies have been made with radiation of low LET, mainly cobalt-60 γ-rays and fast electrons.

It should first be noticed that on grounds of conservation of charge, the net primary yield of hydrogen ions, $G(H_3O^+)$, must equal the primary yield of hydrated electrons:

$$G(H_3O^+) = G(e^-_{aq}) \tag{7.20}$$

Similarly the yields of oxidizing intermediates must correspond on an equivalent basis to the yields of reducing intermediates:

$$2G(H_2O_2) + G(OH) = 2G(H_2) + G(H) + G(e^-_{aq}) \tag{7.21}$$

Of these various quantities, the one which can be measured most directly is the yield of molecular hydrogen, $G(H_2)$. To make this measurement it is necessary to irradiate a solution chosen so that neither H nor e^-_{aq} will give rise to hydrogen. The solute must also react with OH, to prevent decomposition of hydrogen by the reaction:

$$OH + H_2 \longrightarrow H_2O + H \tag{7.22}$$

Aqueous solutions of bromide or many other solutes are suitable. Using suitable solutions it has been found that yields of molecular hydrogen often decrease slightly with increase in solute concentration. This is probably because at high solute concentrations the precursors of the yield of molecular hydrogen are able to react directly with the solute. By plotting the hydrogen yield as a suitable function of solute concentration it is possible to obtain the yield at infinite dilution by extrapolation to zero. A plot of the primary yield of molecular hydrogen against the cube root of the concentration is shown for several solutes in Fig. 7.2 [14]. From the results, the 'primary' yield of molecular hydrogen in the low LET irradiation of pure water, $G(H_2)$, is close to 0.45.

The primary yield of molecular hydrogen peroxide can be determined from measurements of the formation of hydrogen peroxide in the irradiation of water containing dissolved oxygen. In this solution, all of the hydrated electrons and H atoms will react with oxygen:

$$e^-_{aq} + O_2 \longrightarrow O_2^- \tag{7.23}$$

$$H + O_2 \longrightarrow HO_2 \tag{7.24}$$

$$HO_2 \rightleftharpoons H^+ + O_2^- \tag{7.25}$$

so long as the oxygen remains in sufficient excess over the hydrogen peroxide formed. The HO_2 radicals (or O_2^-) give H_2O_2 and O_2 by reaction with each other and the OH radicals react with H_2O_2 to form water and HO_2:

$$OH + H_2O_2 \longrightarrow H_2O + HO_2 \tag{7.26}$$

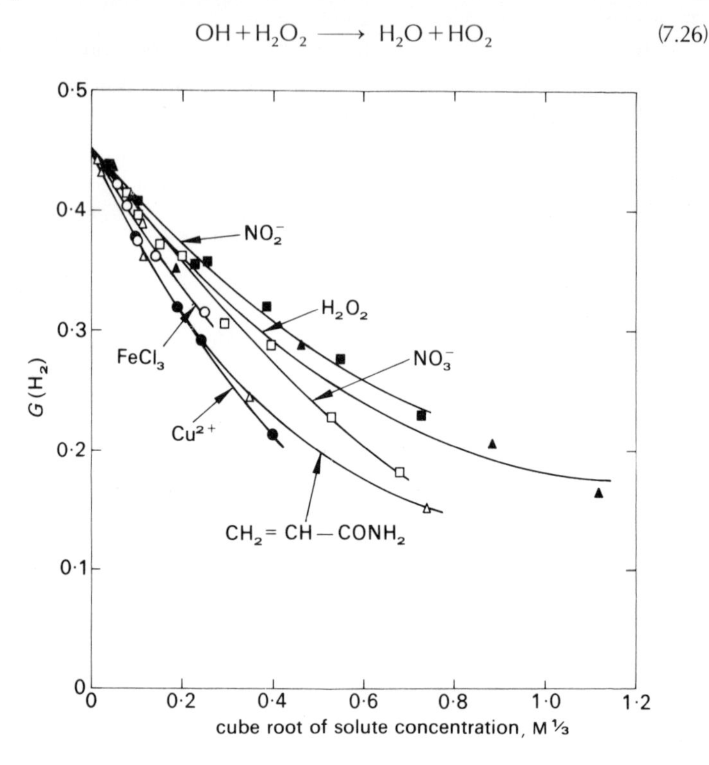

Fig. 7.2 Yield of molecular hydrogen in the low LET irradiation of water as a function of cube root of solute concentration.

Hydroxyl radicals can also react with hydrogen (Reaction 7.22) but if a little bromide is present, the hydroxyl radicals will react with it to give a radical which no longer reacts with hydrogen, though it can react with H_2O_2 as effectively as OH does. From these reactions, the measured yield of hydrogen peroxide in an irradiated solution of oxygen and bromide must be equal to $G(H_2O_2) + \frac{1}{2}(G(e^-_{aq}) + G(H) - G(OH))$, or using Equation 7.21, equal to $2G(H_2O_2) - G(H_2)$. Since $G(H_2)$ is known, $G(H_2O_2)$ can be obtained. Like the yield of primary molecular hydrogen, the yield of molecular hydrogen peroxide often decreases as the solute concentration increases. The yield at infinite dilution can be found by extrapolation. From the irradiation of other aqueous solutions as well as the one discussed, the 'primary' yield of molecular hydrogen peroxide, $G(H_2O_2)$, has been found to be $G = 0.71$ for the low LET irradiation of pure water.

The 'primary' yield of hydrogen atoms may be estimated from systems like the one which originally proved their existence [12]. The hydrogen found after irradiation of an aqueous solution of isopropanol containing acetone will consist only of the primary molecular hydrogen and hydrogen formed by Reaction 7.11. Since the yield of primary molecular hydrogen is $G = 0.45$, $G(H)$ can be found by difference. The conclusion from the low LET irradiation of numerous systems is that $G(H) = 0.55$.

Values of $G(OH)$ and $G(e^-_{aq})$ can be obtained from irradiation of aqueous solutions containing a mild reducing agent, for example, formate, together with dissolved oxygen from the air. Hydroxyl radicals will react with the reducing agent, for example:

$$OH + HCOO^- \longrightarrow H_2O + CO_2^- \tag{7.27}$$

and the radical so formed will react with oxygen:

$$CO_2^- + O_2 \longrightarrow CO_2 + O_2^- \tag{7.28}$$

Both hydrated electrons and H atoms will also give O_2^- (HO_2), which will disproportionate to hydrogen peroxide and oxygen as before. The yield of hydrogen peroxide formed, $G(F)$, can be measured, and will be equal to $G(H_2O_2) + \frac{1}{2}(G(OH) + G(H) + G(e^-_{aq}))$. Using Equation 7.21 we obtain:

$$G(OH) = G(F) - 2G(H_2O_2) + G(H_2) \tag{7.29}$$

$$G(e^-_{aq}) = G(F) - G(H) - G(H_2) \tag{7.30}$$

All quantities on the right hand side of the equations are known, so $G(OH)$ and $G(e^-_{aq})$ can be determined. The best values for the low LET irradiation of pure water are $G(OH) = G(e^-_{aq}) = 2.7$. These and the other yields are summarized in Table 7.1. The values are little if at all dependent on temperature.

Table 7.1 *Yields of intermediates in the radiolysis of pure neutral water with hard X- or γ-rays or fast electrons*

Species	e^-_{aq}	OH	H	H_2	H_2O_2	H_3O^+
G	2.7	2.7	0.55	0.45	0.71	2.7

High concentrations of acid have a small effect on the radical and molecular yields. In solutions of pH 0–1 the total yield of reducing radicals from radiation of low LET becomes 3.7 (it is no longer meaningful to distinguish between hydrogen atoms and hydrated electrons, since the hydrated electrons react according to Reaction 7.10). The yield of molecular hydrogen peroxide goes up to $G = 0.8$ and the yield of OH to $G = 2.9$. The molecular yield of hydrogen is little changed. In alkaline solution yields also appear to increase somewhat. These effects are doubtless basically due to reactions of the precursors of the 'primary products': for example, in neutral solutions there may be a certain reaction of hydrated electrons with hydrogen peroxide within the spurs, but when a high concentration of acid is present, the hydrated electrons in the spurs may react to give hydrogen atoms which are known to react very much more slowly with hydrogen peroxide than hydrated electrons do.

Attempts to measure primary yields in radiolysis with radiation of high LET (for example, polonium α-particles) have given results which are much more dependent on the nature of the solute and its concentration than in radiolysis with radiation of low LET. This is probably because the tracks of high LET particles are so short that the radicals may be considered to be initially disposed in cylinders rather than in isolated groups (Fig. 2.8, p. 31). Consequently, as the radicals diffuse away from the track there is appreciable competition between reaction with other primary products and reaction with solute, whereas with low LET radiation the radicals are very unlikely to react with other radicals after they have diffused even a very short distance away from their point of origin.

One consequence of the increased interaction between 'primary' products in high LET irradiation is that some of the OH radicals must necessarily react with hydrogen peroxide (Reaction 7.26). This reaction may take place between OH radicals and the molecular hydrogen peroxide in low LET irradiation too, but in that case, providing other solutes are present which can react with OH, the probability is so small that the reaction can be neglected. But with high LET irradiation the reaction cannot be eliminated, and HO_2 (or O_2^-) may be regarded almost as a 'primary' product. With polonium α-particles, which have been the most intensively studied of the high LET radiations, the yield is in the region of $G = 0.25$. If HO_2 is regarded as a 'primary' product, Equation 7.21 becomes:

$$2G(H_2O_2) + G(OH) + 3G(HO_2) = 2G(H_2) + G(H) + G(e^-_{aq}) \qquad (7.31)$$

For polonium α-particles the primary yield of molecular hydrogen is about $G = 1.6$–1.8, of molecular hydrogen peroxide 1.1–1.7, and of OH radicals 0.5–0.7. Most of the work with α-particles was done before the discovery of the hydrated electron, and in any case was done with acid solutions: the total yield of reducing radicals is about $G = 0.4$–0.5 [15].

Properties and reactions of the intermediates

Once the 'primary' species (right hand side of Equation 7.12) have been introduced into a solution by means of radiation, they will begin to react chemically. Molecular hydrogen, hydrogen peroxide, and hydrogen ions are relatively inert, but hydrated electrons, hydroxyl radicals and hydrogen atoms are highly reactive. For each of the reactive species, the fraction reacting with a given reactant is equal to the product of the rate constant for the reaction and the reactant concentration, divided by the sum of the products of the rate constants and reactant concentrations for all the other reactions of the species. As the irradiation proceeds, the concentration of the solutes originally present will diminish, and that of the reaction products will increase. The concentration of molecular hydrogen and hydrogen peroxide will build up. The reactions undergone by the primary species will change accordingly.

When the dose-rate is low and the solute concentration not too small, the concentration of hydrated electrons, hydroxyl radicals and hydrogen atoms will be kept to a low level, and there will be little reaction of these primary species with each other. At high dose-rates (as with an electron accelerator) the concentration of primary species can be high, so that interactions between primary species become more important (Reactions 7.14–7.19).

The radiation chemistry of water and dilute aqueous solutions thus reduces essentially to the chemistry of hydrated electrons, hydroxyl radicals and hydrogen atoms. These species will now be considered in turn.

Hydrated electrons Although they were the last to be discovered, hydrated electrons are perhaps the best understood of the reactive primary species [16]. They can be produced without using radiation. One way is by photolysis of iodide, ferricyanide, phenols and so on, in aqueous solution. Another way is by reaction of hydrogen atoms with hydroxide ions:

$$H + OH^- \longrightarrow e^-_{aq} + H_2O \qquad (7.32)$$

They also appear to be formed during the reaction of metallic sodium with water, and during photo-induced electron emission from metallic cathodes. Irradiation however is the most convenient way of producing them.

Hydrated electrons possess many features in common with electrons trapped in solids (pp. 112–113) and with the solvated electrons produced when alkali metals are dissolved in liquid ammonia. The hydrated electron may be visualized simply as an electron surrounded by oriented water molecules:

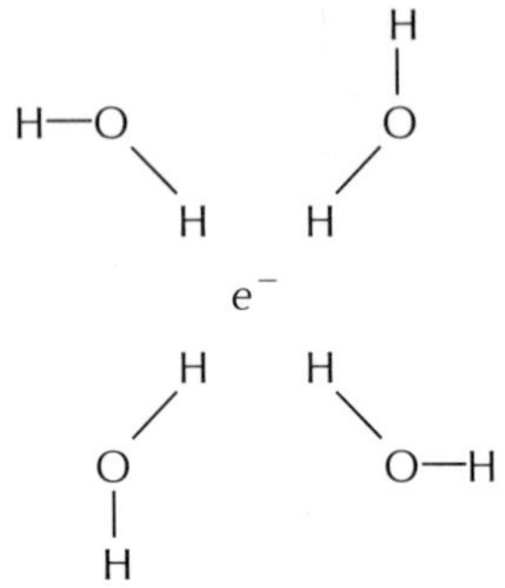

```
                    H
                    |
    H—O             O
       \           /
        H       H
            e⁻
        H       H
       /         \
      O           O—H
      |
      H
```

The structure can be treated by quantum mechanics, and the spectroscopic and kinetic properties calculated. The results are consistent with the mean radius of the charge distribution being 0.25–0.3 nm [17].

The absorption spectrum of the hydrated electron has been given in Fig. 4.11 (p. 92). Owing to the strong broad absorption band it is very easy to look at the reactivity of the hydrated electron by pulse radiolysis. Rate constants are determined by the method already discussed (pp. 90–94). Several hundred rate constants have now been determined for reactions at ambient temperature, more than for any other reactant. A number of activation energies has been determined. Typical rate constants are shown in Table 7.2. Among the reactions of the hydrated electron, that with hydrogen ions is of particular importance because it converts hydrated electrons into hydrogen atoms. Reaction with nitrous oxide is also of special interest, because it gives rise to hydroxyl radicals.

The reaction of the hydrated electron with water (which is extremely slow) may be assumed to give hydrogen atoms and hydroxide ions, though there is no direct proof. The rate constant for this reaction, 16 $\text{M}^{-1}\,\text{s}^{-1}$, may be combined with the rate constant for the reaction of hydrogen atoms with hydroxide ions, which may be taken to be $2.3 \times 10^7\ \text{M}^{-1}\,\text{s}^{-1}$, to give an equilibrium constant for the reaction [18]:

$$e^-_{aq} + H_2O \rightleftharpoons H_{aq} + OH^-_{aq} \qquad (7.33)$$

The value is $16/2.3 \times 10^7 = 7.0 \times 10^{-7}$. It is now possible to attempt to calculate the standard reduction potential for the hydrated electron [19]. We first use the expression:

$$-\Delta G^0 = RT\ln K \qquad (7.34)$$

to obtain ΔG^0 for Reaction 7.33. The answer is 8.4 kcal per mole. The free energy for the hydration of atomic hydrogen is not known, but may be guessed to be similar to that for helium, that is, about 4.5 kcal per mole:

$$H_{aq} \longrightarrow H_{g}, \qquad \Delta G^0 = -4.5 \tag{7.35}$$

Table 7.2 *Rate constants for reactions of the hydrated electron* [20]

Reactant	*Rate constant* ($M^{-1}\ s^{-1}$)	*Reactant*	*Rate constant* ($M^{-1}\ s^{-1}$)
Ag^+	4.0×10^{10}	acetate ion	$<10^6$
Al^{3+}	2.0×10^9	acetic acid	1.8×10^8
BrO_3^-	2.1×10^9	acetone	5.9×10^9
CO	1.0×10^9	acrylamide	3.0×10^{10}
CO_2	7.7×10^9	adenine	9.0×10^9
HCO_3^-	$<10^6$	adenosine-5′-phosphate	4.0×10^9
CO_3^{2-}	$<10^6$	benzene	1.2×10^7
CNS^-	$<10^6$	benzoate ion	3.0×10^9
Cd^{2+}	5.6×10^{10}	benzoquinone	2.7×10^{10}
Cl^-	$<10^4$	carbon tetrachloride	3.0×10^{10}
Co^{2+}	1.3×10^9	chloroform	3.0×10^{10}
Cr^{2+}	4.2×10^{10}	cysteine (pH 6.3)	8.7×10^9
CrO_4^{2-}	1.8×10^{10}	cysteine (pH 11.6)	7.5×10^7
Cu^{2+}	3.0×10^{10}	cystine (pH 6.1)	1.3×10^{10}
e^-_{aq}	$-d[e^-_{aq}]/dt =$	cytosine	1.0×10^{10}
	$1.0 \times 10^{10}[e^-_{aq}]^2$	ethanol	$<10^5$
Fe^{2+}	1.2×10^8	ethylene	$<2.5 \times 10^6$
$Fe(CN)_6^{3-}$	3.0×10^9	formaldehyde	$<10^7$
H	2.5×10^{10}	glucose	$<10^6$
H_3O^+	2.3×10^{10}	glycine (pH 6.4)	8.3×10^6
H_2O	1.6×10^1	guanidine	2.5×10^8
H_2O_2	1.2×10^{10}	histidine (pH 7)	6.0×10^7
I_2	5.1×10^{10}	methacrylate ion	8.4×10^9
N_2O	8.7×10^9	methane	$<10^7$
NO_2^-	4.5×10^9	methylene blue	2.5×10^{10}
NO_3^-	1.1×10^{10}	methyl iodide	1.7×10^{10}
Na^+	$<10^5$	NAD^+	2.5×10^{10}
OH	3.0×10^{10}	oxalate ion	$<10^7$
O_2	1.9×10^{10}	ribonuclease	1.3×10^{10}
$H_2PO_4^-$	4.2×10^6	ribose	$<10^7$
SO_4^{2-}	$<10^6$	styrene	1.5×10^{10}
Tl^+	3.0×10^{10}	thymine	1.7×10^{10}
Zn^{2+}	1.5×10^9	tryptophan	3.0×10^8
acetaldehyde	3.5×10^9	urea	3.0×10^5

where the concentration of hydrogen atoms in the liquid phase is 1M and in the gas phase 1 atmosphere. The free energy change for the reaction:

$$H_g \longrightarrow \tfrac{1}{2}H_{2g} \qquad (7.36)$$

in the gas phase has been established to be $\Delta G^0 = -48.6$ kcal per mole (the concentrations being expressed in atmospheres). For the reaction:

$$H^+_{aq} + OH^-_{aq} \rightleftharpoons H_2O \qquad (7.37)$$

we have $K = 55/10^{-14} = 5.5 \times 10^{15}$, from which $\Delta G^0 = -21.3$. Adding together the Equations 7.33 and 7.35–7.37, we obtain:

$$e^-_{aq} + H^+_{aq} \longrightarrow \tfrac{1}{2}H_{2g}, \qquad \Delta G^0 = -66.0 \qquad (7.38)$$

from which the standard reduction potential for $aq + e^- \rightleftharpoons e^-_{aq}$ is -2.9 volts.†

The reactivity of the hydrated electron can be understood in the light of its thermodynamic properties. It is a powerful reducing agent, and is capable of reducing substances of low standard potential. Thus it can reduce the silver ion although it cannot reduce the potassium ion. In a related sense hydrated electrons are generally captured by substances which are known, or in the absence of experimental data, expected, to have a positive electron affinity. The *rates* of reaction vary from as low as 16 $M^{-1}\,s^{-1}$ to the rates which would be expected if reaction occurred at every collision, as calculated using the Debye equation (Equation 4.15, p. 79) taking the radius of the hydrated electron to be 0.25–0.3 nm and using the experimental value for the diffusion constant [21] which may be taken to be $4.8 \times 10^{-5}\,cm^2\,s^{-1}$. Activation energies of hydrated electron reactions are invariably small, even where low reaction rates are found [22]. Hence the explanation of the variable reactivities is to be sought principally in the entropy of activation. This can best be understood in molecular terms. The controlling factor appears to be availability on the acceptor of a suitable vacant orbital. In the case of simple alcohols, ethers and amines, which possess no low-lying vacant orbitals, the reaction rate is low. On the other hand, substances containing unsaturated groups or halogen atoms tend to have high reactivities. The hydrated electron may be regarded as a reactive nucleophile: in the case of aromatic compounds, withdrawal of electrons from the ring by substituents tends to increase the rate of reaction with the hydrated electron [23]. Similarly, in the case of acids, those with a high dissociation constant

† The reader should take care when consulting other published calculations of this quantity, since nearly all previous calculations contain elementary errors of thermodynamics or arithmetic.

react most rapidly with hydrated electrons perhaps because a high dissociation constant reflects a fractional positive charge on the hydrogen atom, that is, a fractional electron vacancy.

Hydroxyl radicals Free hydroxyl radicals can be produced by the photo dissociation of hydrogen peroxide:

$$H_2O_2 \xrightarrow{h\nu} 2OH \tag{7.39}$$

and may also be formed in the action of ferrous ions on hydrogen peroxide, in the Haber-Weiss reaction:

$$Fe^{2+} + H_2O_2 \xrightarrow{+H^+} Fe^{3+} + H_2O + OH \tag{7.40}$$

A particularly important route to their formation is the reaction of hydrated electrons with nitrous oxide:

$$e^-_{aq} + N_2O \xrightarrow{+H_2O} OH + OH^- + N_2 \tag{7.41}$$

They exhibit an absorption spectrum extending into the ultraviolet, with a molar extinction coefficient at 260 nm of 370 $\text{M}^{-1}\,\text{cm}^{-1}$. However, many other free radicals, and many stable substances absorb in this region, so the disappearance of hydroxyl radicals cannot be followed by pulse radiolysis nearly as easily as the disappearance of hydrated electrons can. Rate constants for reactions of hydroxyl radicals must therefore be determined indirectly, for example by measuring the rate of appearance of the product of the reaction, or by competition methods. Typical rate constants, based on determinations by a number of different methods [24, 25], are shown in Table 7.3.

The standard reduction potential for $OH + e^- \rightleftharpoons OH^-$ is given in standard tables as 1.4 volts [26]. OH is a powerful oxidizing agent but in strongly alkaline solutions its reactivity appears to change markedly. This is explained by the reaction of hydroxyl radicals with hydroxide ions according to:

$$OH + OH^- \longrightarrow O^- + H_2O \tag{7.42}$$

Equilibrium is maintained by the reverse reaction:

$$O^- + H_2O \longrightarrow OH + OH^- \tag{7.43}$$

The p*K* of the radical has been determined by examining the oxidation of certain solutes as a function of pH. With ferrocyanide for example, O^- is comparatively unreactive, but OH reacts rapidly according to:

$$OH + Fe(CN)_6^{4-} \longrightarrow OH^- + Fe(CN)_6^{3-} \tag{7.44}$$

The p*K* determined using ferrocyanide [27] and other solutes is 11.9.

With organic compounds which do not contain double bonds, the commonest reaction of OH radicals is hydrogen atom abstraction, for example:

$$OH + CH_3CH_2OH \longrightarrow CH_3\dot{C}HOH + H_2O \qquad (7.45)$$

OH radicals add to double bonds. With aromatic compounds, they behave like an electrophile, although because of their high electron affinity they appear to be somewhat unselective.

Table 7.3 *Rate constants for reactions of the hydroxyl radical*

Reactant	*Rate constant* ($M^{-1}\,s^{-1}$)	*Reactant*	*Rate constant* ($M^{-1}\,s^{-1}$)
Br^- (pH 7)	1.0×10^9	*n*-butanol	4.2×10^9
CNS^-	1.0×10^{10}	*t*-butanol	5.2×10^8
CO_3^{2-}	3.0×10^8	cysteine (pH 1)	1.2×10^{10}
Fe^{2+}	2.3×10^8	cystine (pH 2)	5.6×10^9
$Fe(CN)_6^{4-}$	9.3×10^9	cytosine	4.6×10^9
H	7.0×10^9	ethanol	1.9×10^9
H_2	4.5×10^7	ethylene glycol	1.5×10^9
HO_2	1.2×10^{10}	formate ion	3.5×10^9
H_2O_2	3.0×10^7	glycerol	1.8×10^9
HSO_4^-	3.0×10^5	glycine (pH 6)	1.6×10^7
OH	$-d[OH]/dt =$	methane	1.4×10^8
	$1.2 \times 10^{10}[OH]^2$	methanol	9.0×10^8
Tl^+	7.6×10^9	*p*-nitrosodimethylaniline	1.3×10^{10}
acetate ion	9.0×10^7	phenylalanine	6.4×10^9
acetone	1.0×10^8	*n*-propanol	2.8×10^9
acetophenone	5.2×10^9	*i*-propanol	2.2×10^9
acrylamide	5.8×10^9	ribose	1.5×10^9
adenine	4.7×10^9	thymine	5.4×10^9
alanine (pH 6)	7.2×10^7	tryptophan	1.2×10^{10}
benzene	7.0×10^9	uracil	5.5×10^9
benzoate ion	5.2×10^9		

Hydrogen atoms These can be produced in aqueous solution without using radiation by bubbling with hydrogen gas through which an electrical discharge has been passed. Some typical constants for their reactions, based on numerous determinations by ESR and other methods, are given in Table 7.4. The usual value given for the standard reduction potential for $H^+ + e^- \rightleftharpoons H$ is -2.1 volts [28]. This value is based on ΔG^0 for Reaction 7.36. Allowance for Reaction 7.35 gives the improved value

-2.3. From the equilibrium constant for Reaction 7.33 the pK of the hydrogen atom is 9.6.

Hydrogen atoms are less powerful reducing agents than are hydrated electrons. In many respects their reactions are closer to those of hydroxyl radicals, although they are distinctly less reactive. With organic compounds containing no double bonds for example, they often abstract hydrogen:

$$H + CH_3CH_2OH \longrightarrow CH_3\dot{C}HOH + H_2 \qquad (7.46)$$

This reaction may be compared with Reaction 7.45. The driving force is the high bond strength of H—H, which is 104 kcal per mole as compared with 120 kcal per mole for H_2O. Hydrogen atoms act like an electrophile in adding on to aromatic rings, and appear to be more selective than OH.

Table 7.4 *Rate constants for reactions of the hydrogen atom*

Reactant	*Rate constant* ($M^{-1}\ s^{-1}$)	*Reactant*	*Rate constant* ($M^{-1}\ s^{-1}$)
Ag^+	1.1×10^{10}	benzonitrile	6.4×10^8
Fe^{2+}	1.8×10^7	*n*-butanol	3.8×10^7
Fe^{3+} (pH 2)	9.5×10^7	cysteine	3.0×10^9
$Fe(CN)_6^{3-}$	4.0×10^9	cystine	6.0×10^9
H	$-d[H]/dt = 2.6 \times 10^{10}[H]^2$	ethanol	2.5×10^7
H_2O_2	6.0×10^7	formate ion	2.3×10^8
NO_3^-	1.2×10^7	formic acid	1.0×10^6
O_2	2.0×10^{10}	glucose	4.0×10^7
OH^-	2.3×10^7	methanol	1.6×10^6
acetamide	1.6×10^5	nitrobenzene	5.6×10^9
acetate ion	2.7×10^5	oxalic acid	3.5×10^5
acetic acid	1.0×10^5	phenol	1.8×10^9
acetophenone	1.2×10^9	*n*-propanol	2.6×10^7
adenine	8.0×10^7	*i*-propanol	7.0×10^7
adenosine	1.2×10^8	thymine	3.0×10^8
benzene	1.1×10^9	tryptophan	9.0×10^9
benzoic acid	9.0×10^8	tyrosine	1.0×10^9

Perhydroxyl radicals Since perhydroxyl radicals are produced in tiny yield in the high LET radiolysis of water they too may be regarded as 'primary' intermediates. Perhydroxyl radicals (HO_2) or superoxide anion radicals (O_2^-) are also formed when hydrated electrons or hydrogen atoms react with oxygen (Reactions 7.23 or 7.24). They can be formed without radiation by the action of certain metal ions on hydrogen peroxide, and can be generated through the photolysis of hydrogen peroxide via Reaction 7.26. The p*K* for the dissociation of HO_2 (Reaction 7.25) is close to 4.9 [29]. Results of one determination of the absorption spectra of the two forms of the radical [29] are shown in Fig. 7.3.

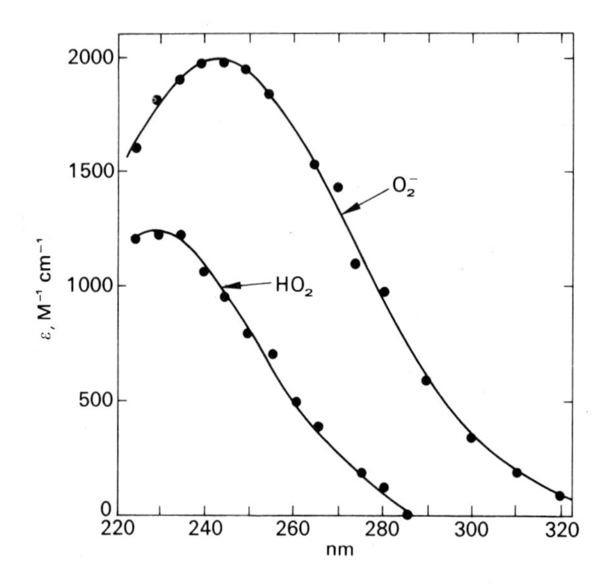

Fig. 7.3 Absorption spectra of the perhydroxyl (HO_2) and superoxide anion (O_2^-) radicals.

HO_2 and O_2^- are both capable of acting as mild oxidizing agents (with the formation of H_2O_2) or mild reducing agents (giving O_2). Unlike e^-_{aq}, OH and H, they are unreactive to most organic compounds. Their fate in irradiated systems often is to disproportionate:

$$2HO_2 \longrightarrow H_2O_2 + O_2 \tag{7.47}$$

$$HO_2 + O_2^- \xrightarrow{+H^+} H_2O_2 + O_2 \tag{7.48}$$

Reactions 7.47 and 7.48 proceed with rate constants $\sim 10^6$ M^{-1} s^{-1} and $\sim 10^8$ M^{-1} s^{-1}, respectively. Superoxide anion radicals appear to react with each other very slowly if at all.

Explanation of the experimental facts in the radiolysis of water

The original experimental work on the radiolysis of water (pp. 136–137) was done before it had come to be accepted that hydrated electrons are major 'primary' products of the radiolysis. Accordingly the earlier explanations of the results [30], although basically correct, now require to be up-dated.

The first step to occur is Reaction 7.12. To account for the stability of pure water under X-irradiation, we have to propose a mechanism by which the H_2 and H_2O_2 can be consumed. It is now known that hydrated electrons react very rapidly with hydrogen peroxide:

$$e^-_{aq} + H_2O_2 \longrightarrow OH + OH^- \qquad (7.49)$$

For X-rays, the primary yield of hydrated electrons is greater than that of hydrogen peroxide, so that in pure water exposed to X-irradiation, the hydrogen peroxide concentration is kept to a very low level by Reaction 7.49. The excess hydrated electrons will react with water (the forward reaction of Equation 7.33) or with hydrogen ions (Reaction 7.10) in either case giving hydrogen atoms. It has long been accepted that hydroxyl radicals will react with molecular hydrogen so that OH, whether formed in the primary process or by Reaction 7.49, can react according to Equation 7.22. This reaction will remove molecular hydrogen. We now have to deal with the H atoms and OH radicals in the system. These interact with each other. If unlike species interact, water is re-formed, as in Equation 7.19. If like species interact, hydrogen or hydrogen peroxide are formed, as in Equations 7.16 and 7.17. The hydrogen and hydrogen peroxide are in turn removed as above.

Now on irradiation with high LET radiation like α-particles, the physical primary events are much closer together, the 'primary' yields of hydrogen and hydrogen peroxide are greater than with low LET radiation, and the 'primary' yields of e^-_{aq}, H and OH are less. Reaction 7.49 still occurs, but to a lesser extent. Hydroxyl radicals can react with hydrogen (Reaction 7.22) but can also react with hydrogen peroxide (Reaction 7.25) at a comparable rate, yielding perhydroxyl radicals. The perhydroxyl radicals (or O_2^-) react to give hydrogen peroxide and oxygen (Reactions 7.47 and 7.48). The oxygen takes up the hydrogen atoms in the system and also some of the electrons (Reaction 7.23) and the radicals formed again give H_2O_2 and oxygen. Comparing high LET irradiation with low LET irradiation, the 'primary' yields of H_2O_2 and H_2, which lead to decomposition, are much greater, while the yields of e^-_{aq}, H and OH, which in low LET irradiation lead to recombination, are much smaller, and the balance of the reactions shifts over towards decomposition.

The enhancing effect of impurities on the low LET-radiolysis of water is due to the reaction of the hydrated electrons and free radicals with the impurities, so hindering decomposition of the molecular products, and possibly also leading to product formation. Where the water contains dissolved oxygen, the hydrated electrons will react with it (Equation 7.23). So will the hydrogen atoms (Reaction 7.24). Both O_2^- and HO_2 will form hydrogen peroxide, via Reactions 7.25, 7.47 and 7.48. Once hydrogen peroxide builds up, the hydroxyl radicals begin to react by Reaction 7.26 and so are no longer effective at removing molecular hydrogen by Reaction 7.22. The enhancement of radiolysis by added hydrogen peroxide is basically due to Reaction 7.26. Hydrogen tends to suppress radiolysis, because it takes up OH radicals (Reaction 7.22) and so prevents them from causing decomposition via Reaction 7.26.

The mechanisms discussed here have been developed in the course of careful experimental work on water containing added hydrogen, hydrogen peroxide and oxygen, and of attempts to put the mechanism on a consistent quantitative basis by estimating the yields of the primary products and the rate constants for the various reactions. Certain experiments done during this work were among those which provided key evidence for the production of hydrated electrons in the irradiation of water [31].

Dilute ferrous sulphate solutions

Dilute solutions of ferrous sulphate or ferrous ammonium sulphate (the ammonium ion plays no part) were among the first relatively simple systems to be irradiated in the hope of understanding the radiation chemistry of aqueous solutions. They have since been very extensively studied. For these reasons, and because of their widespread use for dosimetry (pp. 50–52) they have something of a classic status. Their radiation chemistry will now be discussed to show how the principles discussed so far in this chapter work out when applied to an actual case.

For reasons of stability the ferrous sulphate is usually dissolved in dilute sulphuric acid (0.1–2N). Under ordinary laboratory conditions the solutions contain dissolved oxygen from the air at a concentration of $2\text{–}3 \times 10^{-4}$ M. Most work has been with low LET radiation. Irradiation of the solution converts ferrous to ferric with an initial yield for hard X- or γ-rays or fast electrons of $G = 15.4\text{–}15.6$ which depends very little on ferrous ion concentration in the range 10^{-4} to 10^{-2} M. Initial yields for radiations of various LET are given in Table 3.1 (p. 51). The yield is practically independent of oxygen concentration so long as this is greater than about 5 per cent of the ferrous concentration. When no oxygen is present the yield is smaller, and on prolonged irradiation a steady state is reached which can also be approached by irradiating a solution containing ferric ions and hydrogen. The initial yield is completely independent of dose-rate, but at the high dose-rates attainable with electron accelerators, the amount of ferric formed becomes diminished if the doses are sufficiently great for radicals to react with each other instead of always reacting with solute. Some typical experimental results are shown in Fig. 7.4 [32] and in Table 7.5.

The oxidation originates in the reactions of the intermediates produced in Reaction 7.12. The yields are somewhat greater than those in pure water because of the acidity of the solution, and the 'primary' process may be represented as:

$$H_2O \rightsquigarrow 3.2e^-_{aq} + 2.9OH + 0.5H + 0.4H_2 + 0.8H_2O_2 \quad (7.50)$$

The hydrated electrons react with the hydrogen ions since the rate constant for this reaction is great (Table 7.2) and the concentration of acid is high:

$$e^-_{aq} + H_3O^+ \longrightarrow H + H_2O \quad (7.10)$$

Hydrogen atoms react much more quickly with oxygen than with ferrous (Table 7.4), so that except under extreme conditions such as those applying to the lower part of the second column of Table 7.5, nearly all H atoms,

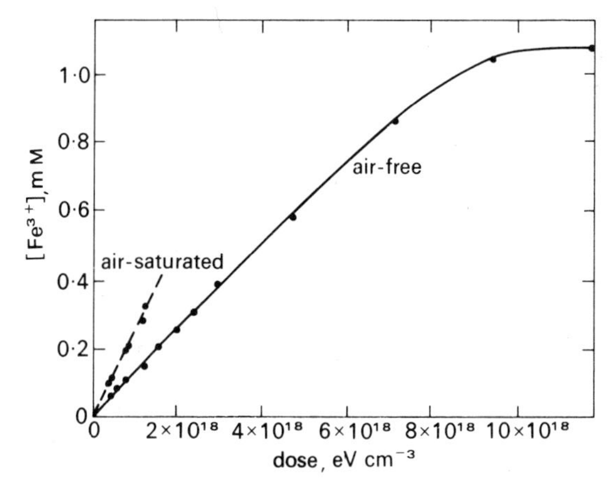

Fig. 7.4 Production of ferric in the X-irradiation of 1.09×10^{-3}M ferrous ammonium sulphate in 0.8N H_2SO_4.

whether produced in Reaction 7.50 or via Reaction 7.10, will react according to:

$$H + O_2 \longrightarrow HO_2 \qquad (7.24)$$

in a solution containing 10^{-4}M oxygen and 10^{-3}M ferrous for example, about 99 per cent of the hydrogen atoms will normally react in this way.

Table 7.5 *Yield of ferric ions in the fast electron irradiation of oxygen-containing solutions of ferrous sulphate in sulphuric acid (0.8N) [33]*

krad per 1.4 μs pulse	*G in air-saturated 10^{-3}M Fe^{2+}*	*G in oxygen-saturated 10^{-2}M Fe^{2+}*
1	15.2	16.0
2	14.9	16.0
4	14.5	15.9
8	13.7	15.8
16	12.6	15.5
32	11.3	15.1
64	9.1	14.5

The HO_2 radicals from Reaction 7.24 react with ferrous to form a hydroperoxide complex, where the HO_2^- may be in the outer sphere [34].

$$HO_2 + Fe^{2+}(H_2O)_6 \longrightarrow Fe^{3+}(H_2O)_6, HO_2^- \qquad (7.51)$$

This species appears to give rise to hydrogen peroxide, so the overall reaction of HO_2 with ferrous may be represented in the simplified form:

$$HO_2 + Fe^{2+} \xrightarrow{+H^+} Fe^{3+} + H_2O_2 \qquad (7.52)$$

Hydrogen peroxide from Reactions 7.50 and 7.52 oxidizes ferrous to give ferric and OH radicals:

$$Fe^{2+} + H_2O_2 \xrightarrow{+H^+} Fe^{3+} + H_2O + OH \qquad (7.40)$$

This is a very slow reaction for radiation chemistry, taking seconds after the hydrogen peroxide has been introduced to become essentially complete. The hydroxyl radicals produced in Reaction 7.40, together with those formed in Reaction 7.50, oxidize more ferrous according to:

$$OH + Fe^{2+} \xrightarrow{H^+} Fe^{3+} + H_2O \qquad (7.53)$$

This reaction may be an outer-sphere electron transfer reaction and so could be written more fully as:

$$OH + Fe^{2+}(H_2O)_6 \longrightarrow \underset{\underset{H_2O}{\downarrow H^+}}{OH^-} + Fe^{3+}(H_2O)_6 \qquad (7.54)$$

Hydrogen atoms can oxidize ferrous and, in the absence of oxygen this is their sole reaction at the beginning of the irradiation. They form an unstable hydride which can be observed by pulse radiolysis and is probably an inner-sphere complex whose formation may be represented simply by:

$$H + Fe^{2+} \longrightarrow Fe^{3+}H^- \qquad (7.55)$$

The hydride reacts with hydrogen ions:

$$Fe^{3+}H^- + H^+ \longrightarrow Fe^{3+} + H_2 \qquad (7.56)$$

When the ferric concentration builds up sufficiently, some of the hydrogen atoms begin to reduce ferric to ferrous:

$$H + Fe^{3+} \longrightarrow H^+ + Fe^{2+} \qquad (7.3)$$

And when the concentration of hydrogen is appreciable compared with the concentration of ferrous, Reaction 7.22 begins to be significant. Reactions 7.3 and 7.22 are the principal reactions responsible for the reverse reaction.

The diminished yields under conditions such as those of Table 7.5, and especially the lower part of the second column, are attributable [33] principally to the occurrence of Reactions 7.16, 7.19 and:

$$H + HO_2 \longrightarrow H_2O_2 \qquad (7.57)$$

in competition for 7.24. Another reaction occurring under these conditions is:

$$OH + HO_2 \longrightarrow H_2O_3 \qquad (7.58)$$

but this reaction as such causes no change in the observed yields since H_2O_3 appears to oxidize ferrous according to the stoichiometry:

$$H_2O_3 + 4Fe^{2+} \xrightarrow{+4H^+} 4Fe^{3+} + 3H_2O \qquad (7.59)$$

Organic compounds produce a marked increase in the amount of ferric formed when aerated solutions of ferrous are irradiated [35]. This is attributable to reaction of OH with the organic compound in competition with Reaction 7.53:

$$OH + RH \longrightarrow R\cdot + H_2O \qquad (7.60)$$

the organic radical forms a peroxide radical by reaction with oxygen and this oxidizes more ferrous than the original OH would have done. The addition of chloride to the solution tends to reduce the effect of organic compounds, since OH reacts with chloride to form a radical which prefers to react univalently with ferrous rather than with organic compounds. It is for this reason that sodium chloride is present in the standard Fricke dosimeter (p. 50).

Other dilute solutions

As well as those already discussed in this chapter, very large numbers of different aqueous solutions have been irradiated. Much of the work has been done to help build up the picture of the radiolysis of water already discussed, while some has been done because of its supposed relevance to radiobiology (see also Chapter 10), nuclear technology, radiation dosimetry and so on. Important work has also been done with the aim of providing insight into the chemistry of substances in unstable valency states (including free radicals) and examples of this work will now be discussed briefly.

The possibility of obtaining such insight arises from the extreme effectiveness of hydrated electrons at causing one-equivalent reductions, whilst OH radicals are very powerful at oxidation and H atoms are also highly reactive. Low LET radiations are used so as to maximize the yields of these species and minimize molecular hydrogen and hydrogen peroxide which are of little interest. Dilute aqueous solutions (about 10^{-5}–10^{-2}M) are now sufficiently understood for it to be sure that there will be no complications due to unsuspected 'primary' processes such as the formation of excited water molecules. Unstable species can most effectively be studied by pulse radiolysis, but low dose-rate irradiation and product analysis also play a part in research in this field.

If it is desired to study a species or a reaction resulting from the action of the hydrated electron it is useful to add to the neutral or alkaline solution an excess of a substance such as sodium formate, methanol or tertiary butanol which will take up OH radicals and in certain cases H atoms, converting them into species which in most cases act as mild reducing agents. To study the action of hydroxyl radicals, solutions may be saturated with nitrous oxide (solubility about 0.025M at atmospheric pressure and room temperature). Provided the solute is not present at too high a concentration, the hydrated electrons will then give further OH radicals (Reaction 7.41) leaving only H atoms whose yield is so small that they can often be neglected. To study species or reactions resulting from the action of H atoms, acid may be added to the solution so as to convert e^-_{aq} into H (Reaction 7.10). Rate constants such as those in Table 7.2–7.4 are used to select appropriate doses as well as compositions of solutions and pH, so as to isolate as far as possible the species or reaction it is desired to study.

One of the earliest studies of this type indicated that hydroxyl radicals add to benzene to form a hydroxycyclohexadienyl radical with an absorption maximum at 313 nm [36], whilst they add to toluene giving a radical (or mixture of isomeric radicals) with a maximum at 317 nm. Benzene itself has a strong peak at 203.5 and toluene at 206.5 nm. By examining numerous aromatic compounds it can be shown experimentally that the wavelength in nm of the observed absorption maximum of the hydroxyl adduct(s) λ_R, is related to that of the parent material, λ_P by the empirical equation:

$$\lambda_R - 1.4\lambda_P = 28 \tag{7.61}$$

a relationship which has proved useful in predicting the position of the maximum of unexamined free radicals [37].

Whereas the addition of OH to benzene is, on a simple view, only likely to give one isomer, addition of H to phenol is likely to give a mixture of different isomers. Accordingly the H- or electron-adduct to phenol give species with an absorption maximum at 330 nm rather than at 313 nm. Presumably the electron adduct becomes rapidly protonated in water. By examining solutions at different pH, different protonated forms of radicals can often be distinguished spectroscopically, and p*K*s determined [37].

The absorption spectra of the H-adduct to benzene is shown in Fig. 7.5 [38]. The position of the absorption maximum is similar to that of the OH adduct, and indeed this is generally found. The wavelengths of the absorption maxima of the H-adducts to benzene and toluene are each about 8–9 nm greater in aqueous solution than in the gas phase.

The organic radicals formed by the action of e^-_{aq}, H or OH often react with each other by dimerization or disproportionation at rates which are somewhat less than diffusion-controlled (about 10^8–10^9 M^{-1} s^{-1}). These reactions are observable in pulse radiolysis. Under low dose-rate conditions, as in γ-radiolysis, each radical must make many collisions with solute molecules (including irradiation products) before it can encounter another radical so that reactions with low rate constants are expected to take place under these conditions.

A beginning has now been made in studying unstable species formed from both organic [39] and inorganic substances. Since the vast majority of chemical substances can be reduced by e^-_{aq} or H or oxidized by OH

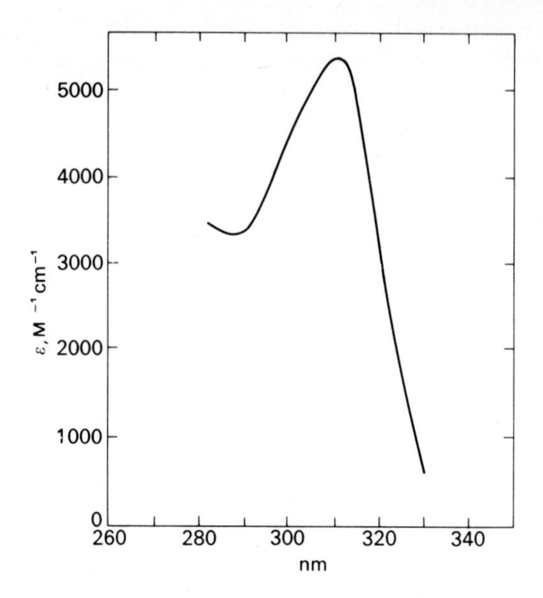

Fig. 7.5 Absorption spectrum of the cyclohexadienyl radical. This is in aqueous solution. In the gas phase the absorption maximum is at 302 nm instead of 311 nm.

or H to give unstable species, and since radiation chemistry and in particular pulse radiolysis are so informative, it is clear that whole new aspects of chemistry are now opened up.

Concentrated solutions

The processes taking place in dilute aqueous solutions must become modified when considering concentrated solutions (greater than about 10^{-2}–10^{-1} M) for several distinct reasons.

First, the rates of the reactions of both 'primary' species (Tables 7.2–7.4) and subsequent reactive species (organic free radicals, metals in unstable valency states and so on) may be appreciably modified by such factors as ionic strength, viscosity and structure of water, all of which may be different in a concentrated aqueous solution from in pure water. When the concentration of reactive solutes is so great that the 'primary' species react before the ionic atmosphere has time to form fully, the dependence on ionic strength is expected to differ from that given by Equation 7.9 [40].

Secondly, the effect of the solute concentration (including H_3O^+ as a solute) on the 'primary' radical and molecular yields is no longer a mere correction to be applied but becomes a dominant process in its own right. If the solute modifies the primary yields it may itself be modified in the process. There are several ways in which this can happen. One is by scavenging of those e^-_{aq}, H and OH which according to the radical diffusion model (pp. 82–83), would otherwise react with each other in dense regions of the particle tracks. Another possibility is that the concentration of solute may be so high that H_2O^+ can react with it (a process which may be called positive hole trapping) before there is time for OH radicals and H_3O^+ to be formed by Reaction 7.13. Alternatively Reaction 7.13 may occur, but H_3O^+ or OH may react before they become full hydrated. Similarly electrons may react with solute before hydration can occur (referred to as reactions of 'dry' electrons) [41]. Furthermore the excited water molecules produced in the primary activation may now have an opportunity to take part in bimolecular processes with solute molecules (cf. p. 67) instead of always decaying in a fashion which is essentially unimolecular.

Finally, when the concentration of solute is more than a few percent by weight, it becomes necessary to consider the direct action of the radiation on the solute. It is not possible to calculate with certainty the fraction of the absorbed energy which is expended in activating each component of the solution, especially when water of hydration is involved. Neither is there any way to predict for sure the chemical behaviour of any solute molecule or ion which becomes directly activated in this way.

Several of the processes which are characteristic of concentrated aqueous solutions may take place at the same time, so it is not easy to establish the extent to which each one contributes.

PROBLEMS

1. By re-plotting the data of Fig. 4.11 with energy in eV as abscissa, estimate the oscillator strength corresponding to the broad absorption shown.

[Ans: 0.7]

2. If the rate constant for the reaction between the hydrated electron and a species M^{2+} is 1.0×10^{10} $\mathrm{M^{-1}\,s^{-1}}$ at zero ionic strength, what would be the rate in a solution whose ionic strength is that of a 10^{-2} M sodium perchlorate solution?

[Ans: 6.8×10^9 $\mathrm{M^{-1}\,s^{-1}}$]

3. In the cobalt-60 γ-irradiation of a 3×10^{-3}M solution of isopropanol containing 10^{-2}M deuteroisopropanol (pH 1) the measured yield of H_2 is 3.03 and of HD is 1.12. What is the ratio of the rate constant for reaction of H with deuteroisopropanol to that for isopropanol if the 'primary' yield of molecular hydrogen in the solution is taken to be $G = 0.43$ and if the reaction of hydrogen atoms with H- containing impurity in the deutero-isopropanol used gives rise to H_2 in a yield which is 7 per cent of the observed yield of HD?

[Ans: see Reference 42]

4. What yield of ferric is implied by the mechanism discussed for the Fricke dosimeter, based on the yields of Equation 7.50? What initial yield would be expected in the absence of oxygen? What is the initial yield in the absence of oxygen in the experiments whose results are given in Fig. 7.4?

[Ans: $G = 15.6$, $G = 8.2$, $G = 8.2$]

5. Consider the pulse radiolysis of a nitrous-oxide saturated aqueous solution of a solute *S*. (*a*) If the rate constant for reaction of e^-_{aq} with *S* is $k = 1.0 \times 10^{10}$ M^{-1} s^{-1}, at what concentration of *S* will 95 per cent of the hydrated electrons react with N_2O? (assume the concentration of N_2O to be 2.5×10^{-2}M). (*b*) At what pH would the number of hydrated electrons reacting with hydrogen ions equal the number reacting with *S* in this solution? (*c*) If the rate constant for reaction of OH radicals with *S* is 1.0×10^9 M^{-1} s^{-1}, at what initial concentration of OH radicals in this solution is the initial rate of reaction of OH radicals with each other equal to one tenth of their initial rate of reaction with *S*? (*d*) From the yields in Table 7.1, what dose in eV cm^{-3} would produce this concentration? (e) What would be the half-life of the OH radical with respect to reaction with *S*?

[Ans: (*a*) 1.1×10^{-3}M (*b*) 3.5 (*c*) 5.8×10^{-6}M (*d*) 1.3×10^{17} (e) 1.0 μs]

REFERENCES

1. I. G. Draganić and Z. D. Draganić. *The Radiation Chemistry of Water*, Academic Press, New York, N.Y., 1971.
2. O. Risse. *Strahlentherapie*, **34**, 578–81 (1929), 'Einige Bemerkungen zum Mechanismus chemischer Röntgenreaktionen in wässrigen Lösungen'.
3. A. O. Allen. *J. Phys. and Colloid Chem.*, **52**, 479–90 (1948), 'Radiation Chemistry of Aqueous Solutions'.
4. H. Fricke and S. Morse. *Phil. Mag.*, **7**, 129–41 (1929), 'The action of X-rays on ferrous sulphate solutions'.
5. J. Weiss. *Nature*, **153**, 748–50 (1944), 'Radiochemistry of aqueous solutions'.

6. A. O. Allen. *Discussions Faraday Soc.*, **12**, 79–87 (1952), 'Mechanism of decomposition of water by ionizing radiations'.
7. G. Stein. *Discussions Faraday Soc.*, **12**, 227–34 (1952), 'Some aspects of the radiation chemistry of organic solutes'.
8. J. H. Baxendale and G. Hughes. *Z. Physik. Chem. (Frankfurt)*, **14**, 306–22 (1958), 'The X-irradiation of aqueous methanol solutions. Part I. Reactions in H_2O'.
9. E. Hayon and J. Weiss. *U.N. Intern. Conf. Peaceful Uses Atomic Energy 2nd Geneva*, 1958, **29**, 80–5, 'The action of ionising radiations (200 kV X rays) on aqueous solutions of mono- and trichloroacetic acid'.
10. G. Czapski and H. A. Schwarz. *J. Phys. Chem.*, **66**, 471–4 (1962), 'The nature of the reducing radical in water radiolysis'.
11. E. Collinson, F. S. Dainton, D. R. Smith and S. Tazuké. *Proc. Chem. Soc.*, 140–1 (1962), 'Evidence for the unit negative charge on the "hydrogen atom" formed by the action of ionising radiation on aqueous systems'.
12. J. T. Allan and G. Scholes. *Nature*, **187**, 218–20 (1960), 'Effects of pH and the nature of the primary species in the radiolysis of aqueous solutions'.
13. A. Hummel and A. O. Allen. *Radiat. Res.*, **17**, 302–11 (1962), 'Radiation chemistry of aqueous solutions of ethanol and the nature of the oxidizing radical OH'.
14. A. O. Allen. *The Radiation Chemistry of Water and Aqueous Solutions.* Van Nostrand, Princeton, N.J., 1961, p. 62.
15. J. Pucheault, in *Actions Chimiques et Biologiques des Radiations*, cinquième série, ed. M. Haïssinsky. Masson, Paris, France, 1961, pp. 31–84, 'Action des rayons alpha sur les solutions aqueuses'.
16. E. J. Hart and M. Anbar. *The Hydrated Electron.* Wiley-Interscience, New York, N.Y., 1970.
17. J. Jortner. *Radiat. Res.*, **Suppl. 4**, 24–34 (1964), 'Addendum: a conjecture on electron binding in aqueous solutions'.
18. E. J. Hart, S. Gordon and E. M. Fielden. *J. Phys. Chem.*, **70**, 150–6 (1966) 'Reaction of the hydrated electron with water'.
19. J. H. Baxendale. *Radiat. Res.*, **Suppl. 4**, 139–40 (1964), 'Addendum: redox potential and hydration energy of the hydrated electron'.
20. Based on M. Anbar, M. Bambenek and A. B. Ross. *'Selected specific rates of reactions of transients from water in aqueous solution. I. Hydrated electron'*, *NSRDS-NBS* 43, Washington, D.C., in the press.
21. K. H. Schmidt and W. L. Buck. *Science*, **151**, 70–1 (1966), 'Mobility of the hydrated electron'.
22. B. Cercek. *Nature*, **223**, 491–2 (1969), 'Activation energies for reactions of the hydrated electron'.
23. M. Anbar and E. J. Hart. *J. Am. Chem. Soc.*, **86**, 5633–7 (1964), 'The reactivity of aromatic compounds towards hydrated electrons'.
24. M. Anbar and P. Neta. *Int. J. App. Radiat. Isotopes*, **18**, 493–523 (1967), 'A compilation of specific bimolecular rate constants for the reactions of hydrated electrons, hydrogen atoms and hydroxyl radicals with inorganic and organic compounds in aqueous solution'.
25. R. L. Willson, C. L. Greenstock, G. E. Adams, R. Wageman and L. M. Dorfman. *Int. J. Radiat. Phys. Chem.*, **3**, 211–20 (1971), 'The standardization of hydroxyl radical rate data from radiation chemistry'.
26. *Handbook of Chemistry and Physics*, 52nd Edn, ed. R. C. Weast. The Chemical Rubber Co., Cleveland, Ohio, 1971–72.

27. J. Rabani and M. S. Matheson. *J. Phys. Chem.*, **70**, 761–9 (1966), 'The pulse radiolysis of aqueous solutions of potassium ferrocyanide'.
28. W. M. Latimer. *Oxidation Potentials*, 2nd Edn, Prentice–Hall, Englewood Cliffs, N.J., 1952.
29. D. Behar, G. Czapski, J. Rabani, L. M. Dorfman and H. A. Schwarz. *J. Phys. Chem.*, **74**, 3209–13 (1970), 'The acid dissociation constant and decay kinetics of the perhydroxyl radical'.
30. A. O. Allen. *The Radiation Chemistry of Water and Aqueous Solutions*, Van Nostrand, Princeton, N.J., 1961, pp. 75–98.
31. N. F. Barr and A. O. Allen. *J. Phys. Chem.*, **63**, 928–31 (1959), 'Hydrogen atoms in the radiolysis of water'.
32. A. O. Allen and W. G. Rothschild. *Radiat. Res.*, **7**, 591–602 (1957), 'Studies in the radiolysis of ferrous sulfate solutions. I. Effect of oxygen concentration in 0.8*N* sulfuric acid'.
33. Based on Table IV of K. Sehested, E. Bjergbakke, O. L. Rasmussen and H. Fricke. *J. Chem. Phys.*, **51**, 3159–66 (1969), 'Reactions of H_2O_3 in the pulse-irradiated Fe(II)-O_2 system'.
34. G. G. Jayson, B. J. Parsons and A. J. Swallow. *Int. J. Radiat. Phys. Chem.*, **3**, 345–52 (1971), 'Aspects of the mechanism of action of low L.E.T. radiation on aqueous solutions of ferrous ions'.
35. H. A. Dewhurst. *J. Chem. Phys.*, **19**, 1329 (1951), 'Effect of organic substances on the γ-ray oxidation of ferrous sulfate'.
36. L. M. Dorfman, I. A. Taub and R. E. Bühler. *J. Chem. Phys.*, **36**, 3051–61 (1962), 'Pulse radiolysis studies. I. Transient spectra and reaction-rate constants in irradiated aqueous solutions of benzene'.
37. B. Cercek, E. J. Land and A. J. Swallow, in *Large Radiation Sources for Industrial Processes*, IAEA, Vienna, 1969, pp. 51–66, 'Short-lived species: identification and kinetic measurements'.
38. M. C. Sauer, Jr. and B. Ward. *J. Phys. Chem.*, **71**, 3971–83 (1967), 'The reactions of hydrogen atoms with benzene and toluene studied by pulsed radiolysis: reaction rate constants and transient spectra in the gas phase and aqueous solution'.
39. E. J. Fendler and J. H. Fendler. *Progr. Phys. Org. Chem.*, **7**, 229–335 (1970), 'The application of radiation chemistry to mechanistic studies in organic chemistry'.
40. P. J. Coyle, F. S. Dainton and S. R. Logan. *Proc. Chem. Soc.*, 219 (1964), 'The probable relaxation time of the ionic atmosphere of the hydrated electron'.
41. W. H. Hamill. *J. Phys. Chem.*, **73**, 1341–7 (1969), 'A model for the radiolysis of water'.
42. P. Neta, G. R. Holdren and R. H. Schuler. *J. Phys. Chem.*, **75**, 449–54 (1971), 'On the rate constants for reaction of hydrogen atoms in aqueous solutions'.

Organic compounds

The principles of radiation chemistry as applied to the condensed phases could not have been developed without using organic compounds, whose tremendous variety of properties is reflected in a corresponding variety of responses to radiation. Among the practical reasons for interest in this field are the need to understand the basic processes occurring during the self-decomposition of labelled compounds and during the damage caused by radiation to the organic materials used in the nuclear energy industry. Some of the discoveries may be put to good use in performing practical organic syntheses, a few of which are of industrial interest. Another reason for work in this field is that simple organic compounds provide models for polymers and for the generally more complex compounds which are of biological interest.

Saturated hydrocarbons

More work has been done on the radiation chemistry of the saturated hydrocarbons than of any other type of compound except water. This is because hydrocarbons are non-polar, contain only simple carbon-carbon and carbon-hydrogen bonds, and are of huge industrial importance. The basic processes taking place on irradiation of saturated hydrocarbons will be discussed first. Actually little direct information is available about the neutral excited states. In general the lifetime is probably too short [1] for them to be able to transfer excitation energy to other substances except at very high solute concentrations, and they must normally dissipate their energy in processes which are essentially unimolecular, sometimes giving rise to the formation of free radicals and molecular products. On the other hand a great deal of information is available about the electrons and positive ions.

Ions Much of the information about the charged species derives from studies using the matrix isolation technique. Most of this work has been done with glasses rather than polycrystalline samples, because glasses are more suited to straightforward optical measurements and because solutes are more readily retained in solution on cooling down. Another characteristic of glasses is that their structure is rather similar to that of liquids, so that many of the conclusions drawn from work with one phase can be carried over to the other. 3-Methylpentane forms excellent glasses at low temperatures and has been widely used [2]. 3-Methylpentane glasses are not hard enough to prevent diffusion processes completely, and harder glasses are needed to stop diffusion even more thoroughly. 3-Methylhexane has been used for this purpose.

When a pure 3-methylpentane glass is irradiated with γ-rays at 77°K it gives rise to characteristic optical [3] and electron spin resonance absorptions which are attributable to trapped electrons. The optical absorption spectrum is shown in Fig. 8.1. The yields of trapped electrons obtained by the two techniques are $G \sim 0.8$ and $G \sim 0.5$ respectively. Such yields are much less than the yields of ion pairs found when gases are irradiated ($G \sim 3$–4) but are significantly greater than the yield of free ion pairs in liquid hydrocarbons at room temperature ($G = 0.146$ for 3-methylpentane). It is easy to see why the yield in the solid should be less than in the gas (Chapter 4, p. 77), yet from Equation 4.14 it would be expected that far fewer electrons would escape from positive ions at the low temperature than at room temperature. The explanation is that in the low temperature solid some of the thermalized electrons go into deep traps in the matrix while still in the coulomb field of the positive ions, instead of recombining at once.

Positive ions must be formed in hydrocarbon glasses in the same yield as the electrons, but cannot be studied as easily and directly. In the first instance the positive charges are probably located on the parent molecules, but the nature of the cations is likely to alter with time through various transfer processes and the positive species may appear ultimately in the form of, for example, carbonium ions. Electrons and positive ions recombine thermally or on illumination, with the production of a weak emission of light. Studies of the kinetics of the recombination show that it is largely a first-order process, consistent with the self-annihilation of isolated ion pairs. However, the recombination cannot be described by

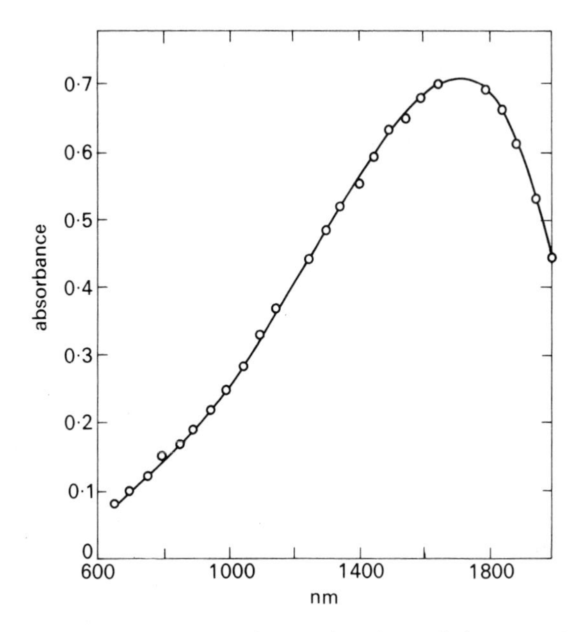

Fig. 8.1 Absorption spectrum of an irradiated 3-methylpentane glass. The absorption is attributed to trapped electrons.

one single first-order rate constant, perhaps because the ion pairs are present in various configurations.

The presence in a hydrocarbon glass of an electron scavenger such as biphenyl enables electrons to be scavenged in competition with physical trapping and ion recombination. Now the biphenylide ion has an absorption maximum at 410 nm. Illumination of the irradiated glass with light of this wavelength detaches the electron from the biphenylide ion, whereupon it becomes trapped in the matrix. Illumination with infrared radiation takes it out of the matrix again, and it attaches to biphenyl once more. This process can be repeated many times with the same sample. Positive holes can migrate through the solid to be trapped on added substances possessing a low ionization potential, such as 2-methylpentene-1, or a high proton affinity, such as methyltetrahydrofuran. Biphenyl, as well as carbon tetrachloride and other substances, can capture positive holes as well as electrons, but the two processes can be distinguished experimentally by judicious choice of additives [4]. When irradiated hydrocarbon glasses containing aromatic substances are warmed up, the positive and negative ions disappear with the formation of solute singlet and triplet states, recognizable by their fluorescence and phosphorescence respectively. Work with naphthalene solutions provides one example of this [5]. Studies of the luminescence emitted from various aromatic compounds enable the recombination processes to be elucidated in some detail [5, 6].

In liquid hydrocarbons, although the yield of free ion pairs is normally about $G = 0.1$–0.2 (Chapter 4, p. 77) and even in exceptional cases is only as high as $G \sim 0.9$, there are several lines of evidence that there is a yield of $G \sim 4$ of positive and negative ions which are capable of entering into specific reactions under suitable conditions. The first evidence was obtained in 1964 from a study of the influence of nitrous oxide on the radiolysis of saturated hydrocarbons [7]. Results are shown in Table 8.1.

Table 8.1 *Effect of nitrous oxide on the γ-radiolysis of cyclohexane* [7]

N_2O concentration (M)	$G(N_2)$	$G(H_2)$
4×10^{-3}	1.14	4.95
2×10^{-2}	2.20	3.80
5×10^{-2}	3.00	3.57
1×10^{-1}	3.75	2.78
2.1×10^{-1}	4.26	2.78

It can be seen that in the presence of nitrous oxide, nitrogen is formed on irradiation and hydrogen (which is a major product in the radiolysis of pure cyclohexane) is produced in a yield which decreases as the nitrous oxide concentration increases. Now nitrous oxide reacts readily with electrons both in the gas phase and in aqueous solution. For liquid cyclohexane it is plausible to suggest that when the concentration of nitrous oxide is high enough, electrons which would otherwise recombine with geminate positive ions may encounter nitrous oxide molecules and react with them to produce the nitrogen:

$$e^- + N_2O \longrightarrow N_2 + O^- \qquad (8.1)$$

Additional reactions of N_2O are not excluded. The reduction in hydrogen yield can be explained if the reaction of an electron with a positive ion gives rise to H_2 but reaction of O^- does not. The results are therefore in accordance with the view that the yield of electrons is of comparable magnitude in the vapour and the liquid phases, and that in the liquid most electrons remain in the field of their positive partners but can react with solutes when the concentration is high. A similar conclusion may be reached from the irradiation of concentrated solutions of ND_3 in C_6H_{12}, in which system the liberated hydrogen gas is found to contain appreciable percentages of HD (Fig. 8.2) [8]. It may be supposed that when the concentration of ND_3 is greater than about 10^{-2}M, the solvent positive ion (which may be either $C_6H_{12}^+$ or $C_6H_{13}^+$ depending on whether the reaction:

$$C_6H_{12}^+ + C_6H_{12} \longrightarrow C_6H_{13}^+ + C_6H_{11}\cdot \quad (8.2)$$

takes place or not) may encounter an ND_3 molecule before neutralization occurs. Since ammonia is basic, it will accept a proton from the positive ion, giving ND_3H^+. When ND_3H^+ ions are neutralized by electrons they will most frequently give ND_2H and D atoms, and the D atoms will abstract H atoms from C_6H_{12} to produce the HD.

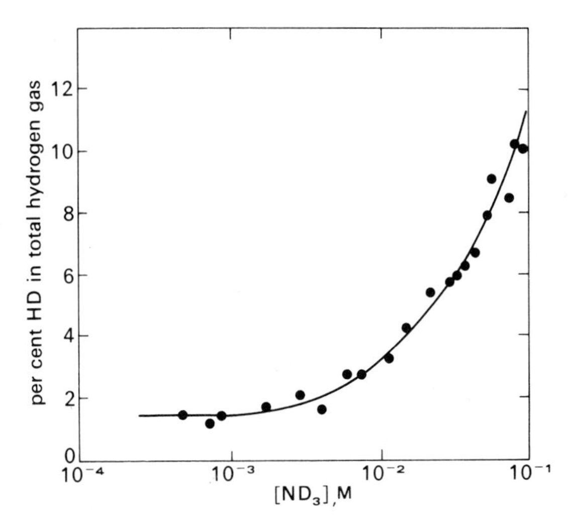

Fig. 8.2 Presence of HD in hydrogen produced by irradiating cyclohexane in the presence of various concentrations of ND_3.

There are numerous solutes besides nitrous oxide and ammonia which can interfere with geminate recombination by capturing electrons or reacting with positive species when present at sufficiently high concentration. Aromatic solutes such as biphenyl, naphthalene, anthracene, pyrene and so on are of special interest since, as already discussed for the hydrocarbon glasses, they may capture charges of either sign and also possess singlet excited states which emit a characteristic fluorescence. Moreover the absorption spectra of their triplet states have been firmly characterized from photochemistry. When a pulse of radiation lasting a few nanoseconds is given to a fairly concentrated solution of one of these substances in cyclohexane (greater than about 10^{-3}M), solute ions can be observed directly. Most of them disappear within about 50 ns, leaving a small number corresponding to the free ions [9]. The characteristic fluorescence of the solute appears, and a high yield of the solute triplet state is found [9, 10]. Similar phenomena occur at low dose-rates, and with other solvents as well as cyclohexane. Whilst various explanations of these phenomena have been discussed, it seems most likely that with saturated hydrocarbons the solute molecules are able to capture either electrons or positive charges, and when the ions of opposite charge come together by diffusion the energy associated with neutralization excites the aromatic solute. This is essentially the same process as takes place in the warming up of irradiated hydrocarbon glasses containing aromatic solutes. Neutralization is expected to produce both singlet and triplet states, the ratio depending on spur size [11]. Some of the singlets may undergo intersystem crossing, producing more triplets.

The kinetics of the reactions of the small numbers of *free* electrons and positive ions in irradiated hydrocarbons are rather simple. Like the hydrated electrons, OH radicals and so on in low LET-irradiated dilute aqueous solutions they may be regarded as essentially dispersed uniformly throughout the medium. They can all react with very low concentrations of suitable solutes. The kinetics of the reactions of the electrons and positive ions which remain within each other's coulomb field are less simple. However, if the electrons which remain within the coulomb field of the positive ions, yield $G(gi)$, are able to react with a solute S to give a product P, the yield of product is expected to follow an equation of the form:

$$G(P) = G(fi) + G(gi)((\alpha[S])^{1/2}/\{1+(\alpha[S])^{1/2}\}) \qquad (8.3)$$

where $G(fi)$ is the yield of free ion pairs and α is a quantity which is proportional to the reactivity of the electrons with the solute [12]. An equation of similar form would apply to positive ion scavenging. When both electrons and positive ions are being scavenged in a solution, the lifetime of the ion pairs will be increased so that except at the extremes of concentrations, the yields of products P would be greater than that given by equations like 8.3.

Free radicals Direct observations of the neutral free radicals formed on radiolysis of saturated hydrocarbons have been made optically and by means of electron spin resonance. The optical absorption spectra are uncharacteristic and give no information about the nature of the radicals, but they permit the reaction kinetics to be studied. The ESR measurements are informative about the radical structure. In the irradiation of pure 3-methylpentane glasses, the ESR evidence shows that a single type of radical, probably $CH_3\dot{C}HCH(CH_3)CH_2CH_3$, appears to be formed with $G = 1.6$ [2]. Hydrogen atoms might be expected too, but are not seen. Thermalized hydrogen atoms would not be able to abstract hydrogen from hydrocarbon molecules at 77°K, so it may be presumed that if hydrogen atoms have been formed at all they have reacted with 3-methylpentane while still hot, or they have reacted with other hydrogen atoms or organic radicals or olefins. No extra radicals are formed on ion recombination so the formation of the radical may be associated with activation to highly excited states.

The formation of a radical from 3-methylpentane by loss of a secondary hydrogen atom is consistent with fission of one of the weak secondary C—H bonds as a result of excitation or ionization. The absence of other types of radical is remarkable, since many different bond rupture processes are known from work in the gas phase. Collisional deactivation and cage recombination of adjacent radicals must play a part in the simplification observed. The formation of a single kind of free radical or at most only a small number of kinds, is quite a general feature in the irradiation of organic solids and liquids, so that γ- or electron-irradiation becomes one of the most useful general methods for the clean preparation of organic radicals for ESR studies [13]. In many cases the relatively small number of kinds of radicals seen is attributable to the more active radicals having undergone chemical reactions with the parent material or the radiolysis products, yielding the less active radicals which are actually observed.

Information about the radicals formed from saturated hydrocarbons in the liquid phase has also been sought by using radical scavengers. Alkyl radicals for example react with iodine to form stoichiometric amounts of alkyl iodides, which can be measured to provide yields of the various radicals produced in the radiolysis. Another method, 'radical sampling', consists of irradiating the hydrocarbon in the presence of a low concentration of either $^{14}C_2H_4$ or $^{14}CH_3I$. Hydrogen atoms will add to labelled ethylene forming $^{14}C_2H_5$ radicals, or electrons will react with the methyl iodide forming $^{14}CH_3$. The labelled alkyl radicals will then react by random combination and disproportionation with all radicals produced in the system. High dose-rates are employed so that radical-radical reactions are favoured over radical abstractions and additions. Analysis of the labelled hydrocarbons formed provides a measure of the nature and yields of the radicals. Before the mid-nineteen sixties it was widely believed that most of the electrons produced on ionization of hydrocarbons would return so quickly to their positive partners that few electrons or positive ions could react with solutes except perhaps at very high concentrations. Solute concentrations as high as about 10^{-2}M were thought to be free from such complications, and were employed for radical scavenging so as to facilitate analytical procedures. However, as discussed above, it is now known that at such concentrations the positive ions and/or electrons may react with solutes. In particular, both positive ions and electrons can react with iodine, while positive ions can react with ethylene. Accordingly the studies intended to measure the yields of the individual types of radicals now have to be interpreted with caution [14]. Nevertheless it is clear that with all liquid hydrocarbons, scission of C—H bonds is an important source of organic radicals. Few radicals are formed by scission of C—C bonds in cyclic hydrocarbons, but in straight or branched chain hydrocarbons, C—C scission is an important source of radicals. Weaker bonds tend to be broken more frequently than the stronger ones, for example the yield of secondary butyl radicals in the radiolysis of liquid *n*-butane is about twice that of *n*-alkyl radicals, although there are six primary C—H bonds in the molecule for every four secondary C—H bonds.

Determinations of *total* radical yields (Table 4.9, p. 82) were in general made with low scavenger concentrations and are probably free from error.

Products Cyclohexane, in which all C—H bonds are essentially identical and all C—C bonds are essentially identical, gives one of the simplest distributions of final irradiation products. The stable products found in pure liquid cyclohexane after γ-radiolysis are listed in Table 8.2. On

Table 8.2 *Products of the γ-radiolysis of cyclohexane* [15]

Product	*Initial G-value*
Hydrogen	5.6 ± 0.1
Cyclohexene	3.2 ± 0.2
Dicyclohexyl	1.76 ± 0.05
1-Hexene	0.40 ± 0.05
Methylcyclopentane	0.15 ± 0.01
Cyclohexylhexene	0.12 ± 0.02
n-Hexane	0.08 ± 0.02
Unidentified C_{12}	~ 0.05
Ethylcyclohexane	~ 0.04
Cyclohexylcyclohexene	0

grounds of material balance, the initial yields of major products should be related by the equation:

$$G(H_2) = G(C_6H_{10}) + G((C_6H_{11})_2) \quad (8.4)$$

1-Hexane and methylcyclopentane do not enter into this equation since their formula is C_6H_{12}, the same as the original material. Taking into account the experimental error, the measured yields very nearly fit the equation, but there appears to be a distinct imbalance. This cannot be accounted for by the minor products, and shows that a small amount of some hydrogen-deficient material must be formed in addition to the products listed in Table 8.2.

The products of the irradiation of liquid cyclohexane are a great deal simpler than those of the irradiation of cyclohexane vapour. In the vapour phase at atmospheric pressure, cyclohexane gives numerous saturated and unsaturated hydrocarbons as significant final products, as well as hydrogen, cyclohexene and dicyclohexyl. The mechanism must involve large numbers of reactions. But at pressures where the density of the vapour begins to approach that in the condensed phase, the yields of fragmentation products are drastically reduced and at 0.42 g cm^{-3} (300°C) for example, the only major products in the γ-radiolysis of cyclohexane are hydrogen ($G = 5.7$), cyclohexene ($G = 1.7$) and dicyclohexyl ($G = 1.1$) [16]. The reason for the differences must be that collisional deactivations and the Franck-Rabinowitch cage effect take place at the high pressures and in the condensed phases, leading to a considerable simplification of products.

With regard to the mechanism of product formation it would seem likely that the cyclohexyl radical would be the immediate precursor to the dicyclohexyl and at least some of the cyclohexene. Positive evidence for the formation of cyclohexyl has been obtained by electron spin resonance measurements on irradiated cyclohexane [17]. The cyclohexyl could be formed from the primary excitations or from the recombination of electrons with positive ions. Several pieces of evidence indicate that the ratio of the rate constants for dimerization and disproportionation of cyclohexyl radicals:

$$2C_6H_{11}^{\bullet} \longrightarrow C_{12}H_{22} \quad (8.5)$$

$$2C_6H_{11}^{\bullet} \longrightarrow C_6H_{10} + C_6H_{12} \quad (8.6)$$

is about 1.1 [18]. If all the dicyclohexyl is formed by Reaction 8.5, the yield of cyclohexene formed by Reaction 8.6 must be $G = 1.6$, and the yield formed by a non-free radical route must be $G = 3.2 - 1.6 = 1.6$. This cyclohexene can be accounted for by decomposition of some of those excited cyclohexane formed in the primary act which do not give cyclohexyl (accompanied by formation of molecular hydrogen or two hydrogen atoms) together with a proportion of the decompositions associated with the energy liberated on neutralization.

The hydrogen may originate in part from molecular detachment of H_2, in part from combination of H atoms with each other:

$$2H \longrightarrow H_2 \tag{8.7}$$

and in part from abstraction reactions of H atoms, especially with cyclohexane itself:

$$H + C_6H_{12} \longrightarrow H_2 + C_6H_{11}^{\cdot} \tag{8.8}$$

Solutes such as iodine, oxygen and benzene decrease the yield of hydrogen from irradiated cyclohexane to $G \sim 3$ when present at sufficient concentration but must be acting by more than one mechanism. Experiments with mixtures of C_6D_{12} and C_6H_{12} provide evidence that at least some of the hydrogen arises from a molecular detachment [19], but it is not easy to assign certain G-values to all the various processes.

The presence of iodine at a concentration of less than about 10^{-4}M has little effect on the yield of hydrogen from the low LET irradiation of cyclohexane [20], although iodine becomes consumed with G about 2.5–2.7 molecules per 100 eV. Under these conditions the electrons and positive ions could account for only a trivial part of the iodine consumption, most of which must be due to the reactions:

$$C_6H_{11}^{\cdot} + I_2 \longrightarrow C_6H_{11}I + I \tag{8.9}$$

$$2I \longrightarrow I_2 \tag{8.10}$$

so that the yield of free cyclohexyl radicals, including those produced by Reaction 8.8, is close to $G = 5.2$. This yield is less than that required to account for the formation of dicyclohexyl with $G = 1.76$ if Reactions 8.5 and 8.6 are both occurring. One possible explanation might be that additional cyclohexyl radicals are giving dicyclohexyl (and cyclohexene) by reaction with each other in spurs before they diffuse into the bulk of the solution where they could react with I_2 if present.

Initial yields in the radiolysis of cyclohexane (Table 8.2) are independent of dose-rate. At low dose-rates many yields become smaller as the irradiation proceeds. At 5×10^{18} eV per cm^3 per hour for example, the yields of hydrogen, cyclohexene and dicyclohexyl at 2×10^{20} eV cm^{-3} are $G = 5.2$, 2.2 and 1.2 respectively [15]. Cyclohexylcyclohexene in contrast is formed with increasing yield as the dose increases, showing that it is a secondary product. By 2×10^{20} eV cm^{-3} the cyclohexene has reached a concentration of 7.3×10^{-3}M, and may react with both H and C_6H_{11}•. A disproportionation reaction of a dicyclohexyl radical would account for the cyclohexylcyclohexene.

Products may enter into other reactions too as their concentration builds up. At higher dose-rates the yields do not decrease as rapidly with dose as at lower dose-rates, presumably because radical-product reactions compete less effectively with radical-radical reactions. If a sufficiently large dose is given in a period which is short compared with the lifetime of the free radicals (for example, with a microsecond electron pulse) the yields of hydrogen, cyclohexene and dicyclohexyl are somewhat less than at low dose-rates, probably because Reaction 8.7 and reaction of H with C_6H_{11}• are favoured at the expense of Reaction 8.8.

Radiations of high LET (for example, 22.9 MeV α-particles, LET = 65 keV/μm^{-1}) produce cyclohexene and dicyclohexyl from cyclohexane with initial yields which are less than for low LET radiation (G = 2.23 and 1.22 respectively for 22.9 MeV α-particles) but the initial yield of hydrogen is not decreased very much, and in fact at LET greater than 10^3 keV μm^{-1} is even greater than for cobalt-60 γ-rays [21]. The deviation from the material balance Equation 8.4 becomes worse the greater the LET. At first sight the reduction in yield of cyclohexene and dicyclohexyl might seem explicable in terms of simple diffusion kinetics (p. 82) since at high LET, as when large doses are given in short times, H atoms would have a good chance of combining with each other as well as of reacting with C_6H_{11}• to re-form cyclohexane, and there would be less of Reaction 8.8. However, the yield of hydrogen would then decrease correspondingly. The fact that the hydrogen yield may even increase, and that unknown products must be present, shows that reactions additional to those discussed must be taking place in the tracks.

When a hydrocarbon contains C—C bonds of different strengths and C—H bonds of different strengths, and does not possess a ring, its radiolysis pattern becomes considerably more complex than that of cyclohexane. This is clear from the experiments with radical scavengers, and can also be seen from the determinations of the formation of products in the radiolysis of pure hydrocarbons. *n*-Hexane may be regarded as typical. In the vapour phase, *n*-hexane, like cyclohexane, undergoes a great deal of both C—C and C—H fragmentation, leading to appreciable yields of hydrogen and saturated and unsaturated hydrocarbons of all molecular weights up to C_{12}. A typical set of results is given in Table 8.3. In the liquid phase the C—C fragmentations are diminished, although not as completely as with cyclohexane. The major products became hydrogen, *trans*-hexene, and various isomeric dodecanes. These are analogous to the main products from cyclohexane. In addition there are significant yields of low molecular weight hydrocarbons and of hydrocarbons with molecular weights intermediate between C_6 and C_{12}.

The following expression of material balance might be expected to apply to the products of irradiation of pure hexane:

$$G(H_2) = \tfrac{1}{2}(-G(C_{lp}) + G(C_{lo}) + G(C_i)) + G(C_d) + G(\text{hexene}) \qquad (8.11)$$

where C_{lp} = low molecular weight paraffin, C_{lo} = low molecular weight olefin, C_i = intermediate molecular weight hydrocarbon, C_d = dodecanes. Application of this equation to the yields in Table 8.3 shows that undetected hydrogen-deficient products must be present. These may include polymeric materials of molecular weight greater than C_{12}.

Table 8.3 *Products of the electron radiolysis of n-hexane* [22]

Product	*G-value (vapour) at about* 1.8×10^{21} eV g^{-1}	*G-value (liquid) at about* 1.8×10^{21} eV g^{-1}
H_2	5.0	5.0
CH_4	0.5	0.12
C_2H_2	0.3	—
C_2H_4	1.1	0.30
C_2H_6	1.0	0.30
C_3H_6	0.3	0.13
C_3H_8	2.3	0.42
C_4H_8	0.06	0.03
i-C_4H_{10}	0.50	0.0
n-C_4H_{10}	2.2	0.50
C_5H_{12}	0.60	0.30
C_6H_{12} (*trans*)	0.10	1.2
i-C_6H_{14}	0.30	0.0
C_7	0.50	0.15
C_8	1.10	0.53
C_9	0.47	0.45
C_{10}	0.14	0.43
C_{11}	0.10	0.02
C_{12}	0.40	2.0

The mechanism of formation of the hydrogen, hexenes and dodecanes must be qualitatively similar to that already discussed for the corresponding products from cyclohexane. The presence of iodine (about 10^{-2}M) depresses the yields of all products from liquid hexane, and in fact the intermediate molecular weight products are almost completely eliminated [23]. At such concentrations iodine would certainly react with all free radicals escaping from the spurs whatever else it may be doing, so these experiments are consistent with some of the hydrogen, hexenes, dodecanes and low molecular weight products being formed by radical reactions outside the spurs, and some being formed by other mechanisms. Intermediate molecular weight products may be formed by combination of radicals with each other in the bulk of the solution.

Carbon-carbon scission giving free radicals may take place both on excitation and ionization of hexane. The radicals could abstract secondary hydrogen from hexane. For example, the *n*-propyl radical may react according to:

$$C_3H_7\cdot + C_6H_{14} \longrightarrow C_3H_8 + C_6H_{13}\cdot \qquad (8.12)$$

or radicals may react with each other by combination, for example:

$$C_3H_7\cdot + R\cdot \longrightarrow C_3H_7R \qquad (8.13)$$

where $R\cdot$ is any of the radicals in the system. Disproportionations may also occur. Reactions like 8.12 will form paraffins of low molecular weight while reactions like 8.13 may be responsible for the formation of the intermediate molecular weight hydrocarbons which are seen. Whether a radical undergoes a reaction like 8.12 or 8.13 must depend on temperature (increasing temperature favouring 8.12 at the expense of 8.13) and dose-rate (high dose-rate favouring 8.13 at the expense of 8.12). γ-Irradiations at temperatures within the range $-78°C$ to $150°C$ do indeed show a decrease in yield of intermediate molecular weight products and increase in paraffins and dodecanes at high temperatures in qualitative agreement with this view. The proportion of dodecanes containing an unbranched C_6 residue decreases with rising temperature [23], consistent with abstractions of secondary hydrogen atoms in reactions of the type 8.12.

As with cyclohexane, another free-radical reaction to be considered is addition of radicals to olefinic products. This will become increasingly important the greater the dose and the higher the temperature.

Despite the undoubted importance of free-radical reactions in the radiolysis of hexane, a high proportion of the low molecular weight hydrocarbon products is probably formed by a different route. One likely path is direct formation of molecules from excited hexane. Another is ion-molecule reactions [24]. The ion $C_4H_9^+$ for example is the most abundant ion in the mass spectrum of hexane and could be formed in the liquid phase too. It could accept a hydride ion from a hexane molecule:

$$C_4H_9^+ + C_6H_{14} \longrightarrow C_4H_{10} + C_6H_{13}^+ \qquad (8.14)$$

giving butane and hexyl ion, which would subsequently be neutralized. Such ion-molecule reactions could be responsible for the formation of an appreciable proportion of several of the products.

As well as cyclohexane and *n*-hexane, large numbers of other hydrocarbons have been irradiated. Overall yields for decomposition of the original material are always about the same. Hydrogen is always one of the most important products, if not *the* most important. Other products include the dimers of parent radicals, together with the saturated and unsaturated hydrocarbons with all molecular weights lower than the dimers. On prolonged irradiation the products are further attacked. One consequence of the further attack is that the viscosity rises owing to the formation of larger and larger molecules. Eventually sufficient molecules join together for a gel to form throughout the system. This happens when lubricants are subjected to very large doses of nuclear irradiation so that materials become converted into crumbly solids. The dose required to form a gel is inversely dependent on molecular weight (Equation 9.30, p. 222) so that gel formation is observed at very much lower doses with polyolefin polymers than with hydrocarbons of relatively low molecular weight. The process is of some technological importance in the polymer field (Chapter 9).

Oxygen, consistent with its reactivity towards electrons and free radicals, exerts a profound effect on the radiolysis of hydrocarbons. For example, when it is present at sufficient concentration during the irradiation of cyclohexane, it reduces the yields of hydrogen, cyclohexene and dicyclohexyl and causes cyclohexanol and cyclohexanone to be formed in approximately equal yield, typically [15] around *G* of 3. Hydroperoxides are formed with a yield of about unity. Above a concentration of around 10^{-3}M, electron capture by oxygen must be significant. Other important reactions must include the reactions of organic radicals and hydrogen atoms with oxygen:

$$C_6H_{11}\cdot + O_2 \longrightarrow C_6H_{11}O_2\cdot \tag{8.15}$$

$$H + O_2 \longrightarrow HO_2 \tag{8.16}$$

followed by disproportionations with oxygen elimination:

$$2C_6H_{11}O_2\cdot \longrightarrow C_6H_{10}O + C_6H_{11}OH + O_2 \tag{8.17}$$

$$C_6H_{11}O_2\cdot + HO_2 \longrightarrow C_6H_{11}OOH + O_2 \tag{8.18}$$

Since the yields at ordinary temperatures are not high it is clear that no chain reaction is occurring. This is consistent with the peroxy radicals not being able to abstract H atoms from the hydrocarbon at the temperature of the experiments. Radiation can induce many other reactions between saturated hydrocarbons and other substances, some of them of value for synthetic purposes [25]. Many of the reactions proceed by free-radical mechanisms. One example, which has been of industrial interest, is the irradiation of hydrocarbons through which gaseous sulphur dioxide and oxygen are passed, leading to the formation of alkylsulphonic acid with yields (under the particular experimental conditions used) of $1\text{–}2 \times 10^3$ molecules of sulphonic acid per 100 eV absorbed [26].

Unsaturated hydrocarbons

Unsaturated hydrocarbons, possessing π-electrons, differ significantly from saturated hydrocarbons in having lower ionization potentials and in being able to accept the addition of free radicals and ions. Also the presence of a double bond in a molecule labilizes any C—H bonds on adjacent carbon atoms (allyl position).

Intermediates 2-Methylpentene-1 glasses, like glasses made from saturated hydrocarbons, give rise to optical [27] and electron spin resonance [28] absorptions attributable to trapped electrons. The position of the optical absorption maximum is close to that shown for 3-methylpentane in Fig. 8.1. From the electron spin resonance spectrum the yield of electrons is $G \sim 0.7$, close to that obtained with the saturated hydrocarbons. An optical absorption band is also seen at 680 nm, and is probably due to some kind of cation. As well as the charged species, the ESR spectrum shows the formation of an allylic radical with $G \sim 0.7$, whose yield and structure could be accounted for [28] if the parent positive ion isomerizes to the cation of 2-methylpentene-2:

$$(CH_2{=}\underset{}{\overset{\overset{CH_3}{|}}{C}}{-}CH_2{-}CH_2{-}CH_3)^+ \longrightarrow (CH_3{-}\overset{\overset{CH_3}{|}}{C}{=}CH{-}CH_2{-}CH_3)^+ \tag{8.19}$$

and then transfers a proton to a neighbouring molecule:

$$(CH_3{-}\overset{\overset{CH_3}{|}}{C}{=}CH{-}CH_2{-}CH_3)^+ + CH_2{=}\overset{\overset{CH_3}{|}}{C}{-}CH_2{-}CH_2{-}CH_3 \longrightarrow$$

$$CH_3{-}\overset{\overset{CH_3}{|}}{C}{=}CH{-}\dot{C}H{-}CH_3 + CH_3{-}\underset{+}{\overset{\overset{CH_3}{|}}{C}}{-}CH_2{-}CH_2{-}CH_3 \tag{8.20}$$

The allyl radical is stabilized by resonance with:

$$CH_3-\overset{\displaystyle CH_3}{\underset{\bullet}{\overset{|}{C}}}-CH{=}CH-CH_3$$

The absence of more radicals can be explained if excited molecules dissipate their energy without giving radicals, or if extra radicals are formed but under the conditions of the experiment disappear in back reactions before they can be detected.

Irradiation of liquid alkenes produce free ions in about the same yield as for the saturated hydrocarbons, in accordance with their similar dielectric constant. Total ion pair yields of about $G = 4$ are to be expected. The main radicals detected in liquid alkenes by ESR and the radical sampling technique are those which correspond to simple loss of a hydrogen atom from the allyl position and those corresponding to addition of a hydrogen atom at the double bond [18]. The mechanism by which the allylic radicals are formed must include proton transfer reactions like 8.20 as well as the direct loss of hydrogen atoms as such as a consequence of the primary activation or H atom attack. Similarly, alkyl radicals would be formed by neutralization by electrons of the carbonium ions formed in reactions like 8.20 as well as by addition of H atoms to the double bond.

Products The effect on the radiolysis products of introducing a double bond into a molecule may be seen by comparing cyclohexene (Table 8.4) with cyclohexane (Table 8.2). It may be seen that the yield of hydrogen is very much less in cyclohexene than in cyclohexane. This is quite general for unsaturated hydrocarbons as compared with the corresponding saturated hydrocarbon. The reason is partly that the π electrons confer some stability on the molecule and partly that any H atoms produced can add on to the double bonds instead of giving rise to hydrogen gas.

Table 8.4 *Products of the γ-radiolysis of cyclohexene* [29]

Product	*Initial G-value*
Hydrogen	1.28
Cyclohexane	0.95
2,2′-Dicyclohexenyl	1.94
3-Cyclohexylcyclohexene	0.60
Dicyclohexyl	0.23
Unidentified dimer	0.22
Polymer (C_6 units)	2.3

The formation of 2,2′-dicyclohexenyl in the radiolysis of cyclohexene may be compared with the formation of dicyclohexyl as a major product in the radiolysis of cyclohexane. It could be formed largely by dimerization of the cyclohexenyl radicals known to be produced as intermediates:

$$2\,C_6H_9\cdot \longrightarrow C_6H_9{-}C_6H_9 \qquad (8.21)$$

Little cyclohexadiene is produced in the radiolysis of cyclohexene [29] so that in contrast to the reaction of cyclohexyl radicals with each other, Reaction 8.21 is not accompanied by any disproportionation analogous to Reaction 8.6. The cyclohexane formed with $G = 0.95$ in the radiolysis of cyclohexene is probably formed partly by disproportionation reactions of cyclohexyl radicals and partly by an ionic mechanism [29]. The 3-cyclohexylcyclohexene could be formed largely by reaction of cyclohexyl with cyclohexenyl.

Table 8.4 shows another general feature in the radiolysis of unsaturated hydrocarbons, that is, that products with two, three or more times the number of carbon atoms in the original molecule are formed in larger yield than from the saturated hydrocarbons. This could arise from the ability of either free radicals or cations to add to double bonds. However, the addition of large alkyl radicals to olefinic double bonds is not a fast reaction at ordinary temperatures. It is in any case in competition with abstraction of allylic hydrogen (where present) to form resonance stabilized radicals as well as with mutual reaction of radicals, so that long free-radical chain additions are disfavoured. Free-radical chain reactions develop only when the double bond is activated, as with the vinyl monomers (Chapter 9). The addition of cations to double bonds under irradiation was first shown with 1-hexene amongst other olefins. The dimeric product was found to contain a high proportion of monolefin, and therefore differs from the dimeric fraction in the radiolysis of cyclohexane and from the diolefin formed by dimerization of allylic radicals in the free radical induced reactions of 1-hexane itself [30]. The principal reaction occurring is probably:

$$R{-}\dot{C}H{-}CH_2^{\,+} + CH_2{=}CH{-}R \longrightarrow R{-}\dot{C}H{-}CH_2{-}CH_2{-}\overset{+}{C}H{-}R \qquad (8.22)$$

followed by an internal hydride ion shift and neutralization by an electron to produce 4-dodecene. Trimers, tetramers and so on are also found as a consequence of further reactions of cations like those formed in Reaction 8.22. Similar reactions may occur in cyclohexene in addition to the free-radical reactions discussed above. The free ions produced by irradiation of olefins can sometimes initiate very long chain reactions, as studied in detail for isobutylene and cyclopentadiene amongst other compounds [31], so that high polymers may be produced (also see Chapter 9).

In contrast to the saturated hydrocarbons, the *G*-values for the formation of products from olefins are independent of dose up to quite large doses. For example, in the radiolysis of cyclohexene the *G*-values shown in Table 8.8 apply up to at least 2×10^{20} eV cm^{-3}. This is because, unlike with the saturated hydrocarbons, the products are in general little more sensitive to attack by intermediates than is the original material. Radiations of high LET produce substantially more hydrogen from unsaturated hydrocarbons than do low LET radiations and, as with the saturated hydrocarbons, this is due to reactions taking place in the particle tracks which do not take place significantly in low LET irradiation [32].

Another important response of olefins to irradiation is to undergo *cis-trans* isomerization. The reaction can take place in high yield when the olefin is irradiated in solution in aliphatic or aromatic hydrocarbons. There are several possible mechanisms including ionic and free-radical reactions, and transfer of energy from solvent excited states. Determination of the yield under carefully chosen conditions has been used to provide a measure of the yield of solvent excited states in the radiolysis of aromatic hydrocarbons (see below, p. 188).

Olefins are much more susceptible to radiolytic oxidation than are paraffins. In the case of cyclohexene for example, peroxide has been found with *G* up to several hundred after X-irradiation in the presence of oxygen at 28–63°C [33]. The reason for the high yield is the ease of abstraction of an allylic hydrogen atom by peroxide radicals, so that the propagating steps are probably:

$$C_6H_9\cdot + O_2 \longrightarrow C_6H_9O_2\cdot \tag{8.23}$$

$$C_6H_9O_2\cdot + C_6H_{10} \longrightarrow C_6H_9O_2H + C_6H_9\cdot \tag{8.24}$$

This is a classical free-radical chain oxidation mechanism. In agreement with the accepted kinetics for such reactions initiated in other ways, the rate of oxidation is proportional to the square root of the dose-rate. The activation energy is 8.1 kcal per mole [33]. Oxidation of olefinic compounds under irradiation has technological importance in a number of fields, for example in damage caused by the irradiation of hydrocarbon polymers in the presence of oxygen (see also p. 217) and in the irradiation of foodstuffs, where the fatty components can oxidize. It is possible for the oxidation to have a longer chain length in the solid phase at low temperature than in the liquid phase. This is because the termination by mutual reaction of two peroxy radicals becomes hindered in a solid since the radicals cannot meet each other so readily by diffusion, whereas the propagating steps involving the small oxygen molecule (Reaction 8.23) or a neighbouring molecule (8.24) have plenty of opportunity to take place [34]. A similar phenomenon increases the rate of polymerization of monomers at high degrees of conversion (p. 207) and during polymerization in the solid phase (p. 209) and in grafting (p. 225).

Dozens of free radical additions to the double bonds of olefins and other unsaturated compounds have been produced by radiation [25]. The addition of HBr to ethylene provides a simple example (p. 132). These reactions may be regarded as extreme examples of chain transfer (p. 207). Some are of potential or actual interest to the chemical industry as a basis for commercial processes.

Aromatic hydrocarbons

Because of the highly conjugated π-electron system, the energy of excitation of aromatic compounds is highly delocalized instead of being connected with particular bonds. Even in the liquid phase, where collisional quenching is always at its strongest, the energy taken up by aromatic compounds can be retained for relatively long times in the form of electronic excitation, and the lifetime of the excited state can, in favourable cases, be as long as nearly one microsecond for certain excited singlets and many seconds for triplets. With certain compounds, known as 'scintillators', the excited states can emit light with great efficiency, and in favourable systems as much as about 10 per cent of the radiation energy absorbed can be converted into light. Another characteristic of aromatic hydrocarbons is that their ionization potentials, like those of olefins, are less than those of saturated hydrocarbons so that the molecules can accept positive charge. Unlike the simple olefins, aromatic hydrocarbons are able to capture electrons, sometimes with high efficiency. They readily accept the addition of free radicals.

Excited states The highly excited singlet states produced in the primary activation of aromatic hydrocarbons are in general expected to undergo internal conversion to lower levels within about 10^{-11} s. Because of their short life, it is difficult to study them directly, although there is theoretical and some experimental evidence consistent with their being able to undergo bimolecular reactions with solutes if the solute concentration is high enough [35]. In the case of benzene, experimental evidence is provided by studies of the effect of chloroform on the luminescence emitted from solutions containing phenylbiphenyl oxadiazole (PBD), a representative scintillator [36]. Now in the absence of chloroform, it is known that benzene molecules excited to the lowest singlet excited state by illumination with ultraviolet light can transfer their energy efficiently to PDB which then efficiently fluoresces. The addition of chloroform makes no difference to the fluorescence induced in this way, showing that the lowest singlet excited state of benzene is not quenched by chloroform. A high degree of fluorescence is also observed on γ-irradiation of PBD in benzene, showing that energy absorbed in benzene becomes transferred to PDB. However, in this case the addition of chloroform (10^{-2}–10^{-1}M) does reduce the luminescence. These observations can be explained if the highly excited benzene molecules can react with chloroform to form a charge-transfer state with benzene as electron donor and chloroform as electron acceptor: this would interfere with internal conversion to the lowest excited state and so lead to a reduced yield of the lowest excited singlet of benzene and hence of PDB and therefore to a smaller amount of light. Determinations of free-radical yields in the system are also consistent with this view. An alternative explanation would be that the chloroform captures an electron emitted from benzene while it remains in the coulomb field of its positive ion, and so prevents electron-positive ion recombination giving rise to the formation of the excited singlet state of benzene. It is difficult to distinguish between these explanations, and indeed the distinction may be simply a matter of semantics.

The lowest excited singlet states of benzene† and other aromatic molecules might be expected to be formed as a consequence of the primary activation through internal conversion from higher levels. Some would also be formed by direct excitation. More would be formed together with triplets during any recombination of ions. Once formed, they may undergo intersystem crossing to the triplet level, or they may react chemically, transfer their energy to solutes, luminesce or dissipate their energy in the form of heat (Chapter 4). The same excited states, reacting in the same

† The most frequently discussed excited states of benzene are often referred to by the following symbols:

Lowest excited singlet state, energy 4.8 eV	$^1B_{2u}$
Excimer corresponding to the $^1B_{2u}$ level, energy 4.4 eV	$^1B_{1g}$
Strongly allowed singlet state, energy 6.9 eV	$^1E_{1u}$
Lowest triplet state, energy 3.7 eV	$^3B_{1u}$

way, may be formed photochemically. The lowest excited singlet state of benzene has been seen by pulse radiolysis in the form of its excimer [37]. The species fluoresces and possesses an absorption maximum at 515 nm. Both fluorescence and absorption decay with a half-life of 18 ns. Measurements of the large amounts of fluorescence from solutions containing scintillators enable further determinations to be made of the lifetime of the benzene singlet (or its excimer) with respect to unimolecular decay, and also of the rate constant for reaction with solute. In a typical experiment, pulses of X-rays were given to solutions of *p*-terphenyl of concentration 4×10^{-4}–2×10^{-3} M in benzene. In this solution the benzene singlet or its excimer may decay unimolecularly with a mean life τ, or may transfer its energy with a second order rate constant k to solute of concentration [S]. The solute then emits fluorescent radiation, lifetime 2 ns. Under the conditions of the experiment, the observed lifetime of the fluorescence, τ^0, equals the lifetimes of the benzene singlet and is governed by the equation:

$$1/\tau^0 = 1/\tau + k[S] \qquad (8.25)$$

In the experiments quoted, τ was found to be 15.2 ns, that is, half-life 10.5 ns, and k was found to be 4.9×10^{10} M^{-1} s^{-1} [38]. Other measurements, including photochemical measurements, have given rather similar values. Rate constants for transfer of singlet excitation energy are very often found to be higher than if they were diffusion-controlled, as given by Equations 4.6 (p. 70) or 4.16 (p. 79), showing that energy-transfer processes other than simple collisions of the second kind are taking place.

The *G*-value for the formation of the lowest excited singlet state of benzene under low LET irradiation has been estimated by measuring the amount of light emitted from an irradiated solution of *p*-terphenyl in benzene, and comparing this with the amount of light emitted when the *p*-terphenyl in an identical solution is directly excited by absorption of light in its own absorption band [39]. The *G*-value found for ^{14}C beta radiation was 1.55. Other determinations have given similar values. *G*-values for alkyl benzenes, stilbene, naphthalene, terphenyl, anthracene and so on are also in the region 1–2.

From experiments in which the lowest excited singlet state of benzene is excited directly by ultraviolet light, several independent determinations show that the crossover efficiency for singlet→triplet is about 0.5.

The primary activation of aromatic compounds would be expected to give rise to triplet states as well as singlets if ion-recombination takes place (cf. p. 79). Another source of triplets would be the intersystem crossing from excited singlets. An unknown but small amount could also be formed by direct excitation by slow electrons (p. 27). The triplets could transfer their energy to suitable solutes at rates which approach, but do not exceed, those calculated for reaction which take place at every collision. They could luminesce (phosphoresce) but, because the lifetime for this process is often long, the luminescence is prone to be quenched through collision with impurities, radicals, radiolysis products and so on. In the liquid phase, triplets are nearly always quenched in this way but in the solid phase, where all species are immobile, the quenching process is much less in evidence. Oxygen is a particularly ubiquitous and effective quencher.

The triplet state of benzene itself has an exceptionally short lifetime in the liquid phase (<10 ns). It has not been observed directly in radiolysis, but information about it has been gained by irradiating benzene solutions of solutes which are expected to accept triplet excitation energy from benzene and then respond in a known way. Olefins such as polybutadiene, 2-butene and stilbene have been used since these compounds are known to undergo isomerization on excitation. Fig. 8.3 shows one of the sets of results obtained in one of the studies on the effect of concentration

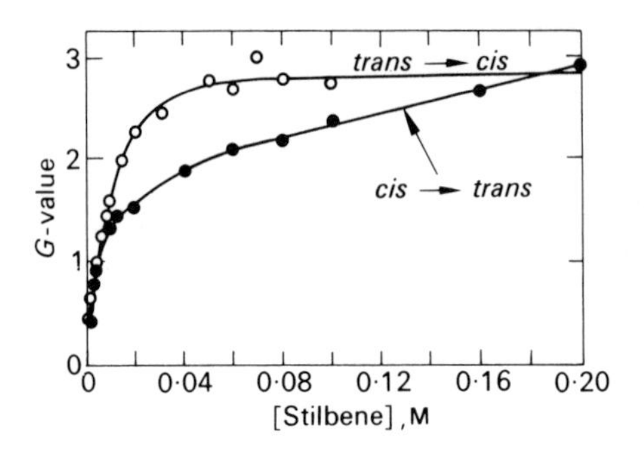

Fig. 8.3 Yield for isomerization of stilbene on irradiation in benzene solution at various concentrations.

on the yield of *trans*-stilbene in the γ-irradiation of solutions containing *cis*-stilbene, and of *cis*-stilbene in the irradiation of solutions containing *trans*-stilbene [40]. The total amount of stilbene remains constant throughout these irradiations, showing that isomerization is the only significant change taking place. Up to a concentration of about 0.1M, the ratio of the *G*-value of *trans* → *cis* and *cis* → *trans* is about 1.2, as in the photochemical isomerization. At higher concentrations, different effects are present, perhaps associated with higher excited states or with ionic reactions produced by interference with geminate ion recombination. The results of Fig. 8.3 may be usefully plotted as the reciprocal *G*-value against reciprocal stilbene concentration (Stern-Volmer plot, pp. 69–70). This procedure gives a good straight line for the results up to 0.1M with, in the case of the *trans* → *cis* isomerization, an intercept at zero reciprocal solute concentration corresponding to $G = 2.9$. This G-value is what the limiting G-value for isomerization would be at infinite solute concentration if the additional effects above 0.1M did not take place. Now stilbene quenches the fluorescence of benzene rather efficiently by accepting energy from benzene excited singlets. The quantum yield of *trans* → *cis* isomerization from this source may be taken to be 0.48 molecules isomerized per stilbene excited singlet state produced [41]. If we take the yield of the lowest excited singlet in the radiolysis of benzene to be $G = 1.5$, the limiting yield for *trans* → *cis* isomerization of stilbene by energy transfer from the lowest excited state of benzene becomes $G = 1.5 \times 0.48 = 0.7$. The remaining portion, $G = 2.9 - 0.7 = 2.2$ comes from benzene triplet. The quantum yield for *trans* → *cis* isomerization from the triplet state may be taken to be 0.55 [40], from which the yield of triplet states in benzene formed without participation of the lowest excited singlet becomes $2.2/0.55 = 4.0$. If no solute were present to remove the excited singlet, an additional yield of $G = 1.5 \times 0.5 = 0.75$ would be formed by intersystem crossing. It is possible to derive somewhat similar values from suitable pulse radiolysis experiments in which solute triplets are observed after the pulse. The addition of solutes as diverse as N_2O, cyclohexene, piperylene and xenon profoundly modifies the process taking place, but the interpretation of the results is rarely as simple as might be thought.

As already mentioned above (pp. 169–172) both singlet and triplet states of aromatic compounds can be formed when the compound is irradiated in solid or liquid solution in suitable non-aromatic solvents. Regardless of the mechanism, the formation of excited states in such ways becomes a useful method of preparing them so that their properties and reactions can be studied.

Ions and free radicals The yield of free ion pairs in the γ-irradiation of liquid benzene is only $G = 0.053$, consistent with most of the initially produced electrons returning rapidly to their positive ion partners. As with saturated hydrocarbons, the γ-radiolysis of aromatic hydrocarbons containing nitrous oxide gives rise to the formation of nitrogen. However, the N_2O concentrations required to produce appreciable yields of nitrogen are an order of magnitude greater than with saturated hydrocarbons. For example, with benzene an N_2O concentration of 0.1*M* is required to produce $G(N_2) = 1.0$. This can easily be understood on the basis that aromatic hydrocarbons capture electrons whereas saturated hydrocarbons do not, so that more N_2O is required to compete for electrons.

The result could also be explained on the basis of reactions of higher excited states similar to those discussed above (p. 186). The basic role of the N_2O would then be to act as acceptor in a donor-acceptor complex.

Although ions have not been detected in appreciable yield in the irradiation of pure aromatic compounds, aromatic anions and cations can be observed on irradiation of solid or liquid solutions of aromatic hydrocarbons in aliphatic hydrocarbons as discussed above. Aromatic anions (or their neutral protonated forms) can also be prepared in aqueous solution by reaction of hydrated electrons with the compound (pp. 150, 160–161) or in alcohol solution similarly. Cations can be prepared in dichloroethane or other chlorinated solvents. Experiments with these solutions provide much information about the properties of the aromatic ions. For instance studies of alcohol solutions containing two aromatic compounds have shown that rates of transfer of electrons from one compound to another approach, but do not exceed, the diffusion-controlled rate [42].

Neutral aromatic free radicals can very easily be observed by pulse radiolysis. Those formed in the radiolysis of pure compounds cannot easily be identified, but it is relatively simple to identify the radicals formed on radiolysis of suitable solutions. The benzyl radical was one of the first radicals to be seen by pulse radiolysis [43]. Aromatic radicals tend to possess characteristic optical absorption bands in easily accessible regions of the spectrum, which facilitates the study of their properties and reactions. Studies in aqueous solution are particularly informative (p. 160), since under those conditions the reactions by which the radicals are produced are quite clear. The ESR spectra of aromatic radicals have been observed after irradiation of solid aromatic compounds. The yields found are quite small, only $G = 0.03–0.3$, much less than for saturated hydrocarbons or non-conjugated olefins [44]. In the case of benzene the principal radical observable is cyclohexadienyl, but the observed spectrum is not simple and other radicals must be present too.

Products The products of the γ-irradiation of liquid benzene at ordinary temperatures are listed in Table 8.5. The 'polymer' is an unsaturated material whose molecular weight becomes greater the greater the dose reaching, for example, a molecular weight of 500 at 200 Mrad. Apart from the polymer, the yields are little dependent on dose. By comparison with the irradiation of the saturated hydrocarbons or the olefins, it can be seen that the product yields are very small, showing the ability of liquid benzene to take up energy without decomposition. The yields in the liquid phase are much less than in the vapour phase. The smallness of the yields in the liquid phase contrasts with the relatively high yields of the lowest singlet and triplet excited states. It is plausible to suggest that the small amount of chemical decomposition in the liquid arises in some way from excited states, but that once the lowest excited states have been reached the molecules cannot efficiently decompose. In agreement with this view, *p*-terphenyl, although efficient at removing excitation energy from the lowest excited singlet state of benzene, produces no effect on the yield of hydrogen when present during irradiation at concentrations up to 0.04M [45].

Table 8.5 *Products of the γ-radiolysis of benzene* [46]

Product	*Initial G-value*
Hydrogen	0.039
Acetylene	0.020
Ethylene	0.022
1,3-Cyclohexadiene	0.008
1,4-Cyclohexadiene	0.021
Phenyl-2,4 cyclohexadiene	0.021
Phenyl-2,5 cyclohexadiene	0.045
Biphenyl	0.065
Polymer (C_6 units)	0.8

Some of the yields from liquid benzene increase somewhat with irradiation temperature in the range up to 150°C. For example the yield of 1,4-cyclohexadiene is more than twice as great at 150° as at 10°. However, the variation with temperature does not follow any simple law. Yields are strongly dependent on LET. For instance, the yield of hydrogen is about three times as high on irradiation with 0.69 MeV protons as on irradiation with γ-rays [47], and even higher values can be calculated for the instantaneous *G*-values in parts of the track where the local rate of energy deposition is at its highest. The effect of LET seems to be interpretable in terms of interactions between excited states (perhaps including higher or charge-separated states) in the particle tracks. The greater the local LET, the more these reactions, which may be represented:

$$2B^* \longrightarrow \text{products} \tag{8.26}$$

would be preferred over first-order deactivations of B^*. However, since benzene appears so resistant to the effects of electronic activation, it also becomes necessary to consider other possible sources of chemical decomposition. It is likely in particular that the knocking-on of protons at the end of the tracks of heavy particles might cause appreciable decomposition [48], and with certain radiations such as epithermal neutrons and with other aromatic compounds if not benzene, this may well become the dominant primary process.

Other aromatic compounds share with benzene the property of being relatively resistant to irradiation. This opens up the prospect of using them for certain technological purposes where other materials would be too susceptible to radiation damage. One such possibility is to use polyphenyls as coolant-moderators in nuclear reactors, since polyphenyls have a long history as heat-transfer fluids in other situations, and would also act as good moderators for fast neutrons [49]. Without radiation, compounds such as biphenyl, *o*-, *m*- and *p*-terphenyl and naphthalene are adequately stable at the temperatures to be encountered in reactors. Under the influence of low LET radiation the materials increase in viscosity owing to the formation of material of high molecular weight, with initial yield for the terphenyls of $G = 0.2–0.3$ molecules consumed per 100 eV. With high LET radiations such as those in a reactor the yields are several times greater and increase markedly with temperature, reaching, for *p*-terphenyl, $G = 1.3$ at 400°C. The high molecular weight material tends to precipitate on heat transfer surfaces, and this is a major obstacle to the economic use of organic-moderated reactors.

Another application where aromatic compounds can be useful is in the formulation of radiation-resistant lubricants. The effects of radiation on lubricants (p. 180) are generally unimportant below 1 Mrad, but above 100 Mrads many lubricants are seriously damaged by the increase in viscosity. The development of acidity, especially when oxygen is present, is also damaging. Doses of 100 Mrads could be reached in a few hours in parts of certain nuclear reactors, so the effect of radiation on lubricants has had to be considered seriously. Very many commercially available and specially formulated oils and greases have now been examined. The nature of the organic base fluid is the most important consideration for radiation stability. Aromatic fluids such as polyphenyls, poly(phenyl ethers) [C_6H_5 (—OC_6H_4—)$_n$ OC_6H_5] and alkyl aromatics are the most stable. Another factor is that radiation also causes serious damage to the additives which are normally present in lubricants [50]. Radiation-resistant oils and greases, suitable for doses up to nearly 10^4 Mrads, have been formulated, and are now commercially available.

Although aromatic compounds are relatively resistant to the direct effect of radiation, their susceptibility to attack by free radicals and electrons enables products to be formed with *G*-values in the region of three on irradiation in solvents such as water. The formation of phenol on irradiation of benzene-water systems and of chlorinated hydrocarbons by a chain reaction in the irradiation of aromatic hydrocarbon-chlorine mixtures are among the reactions to have been investigated as a basis for industrial processes [25].

Alcohols

The properties of the simple alcohols are intermediate between those of water, which contains only O—H bonds, and the saturated hydrocarbons, which contain only C—H bonds and (except for methane) C—C. As just one example the simple alcohols have static dielectric constants at ordinary temperatures of 20–30, compared with 80 for water and 2 for hydrocarbons. In accordance with this status, the features found in the radiation chemistry of the alcohols are in between those found in the radiation chemistry of the other two.

Intermediates As with both water and aliphatic hydrocarbons, there is little direct information about the chemistry of the excited states formed in the primary activation of alcohols. Presumably they disappear very rapidly ($<10^{-12}$ s), perhaps forming radicals and/or molecular products.

Trapped electrons have been observed optically and by electron spin resonance in irradiated alcohol glasses at 77°K. Immediately after a nanosecond pulse, the optical absorption spectrum shown by an ethanol glass appears to be rather like that of an irradiated hydrocarbon glass, but within a few microseconds the shape of the spectrum changes and an absorption maximum appears at about 550 nm (Fig. 8.4) [51]. Probably

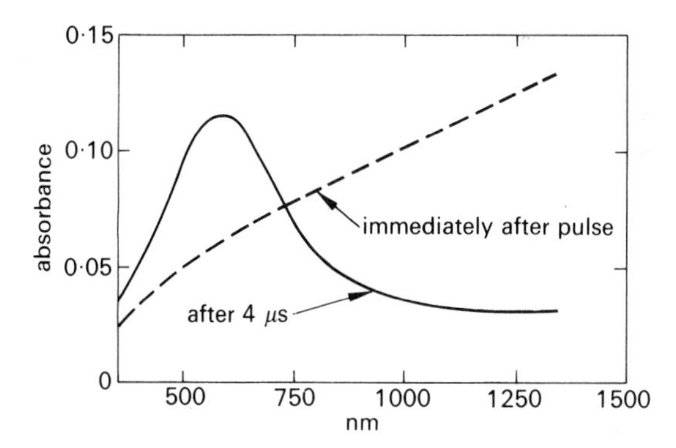

Fig. 8.4 Absorption spectra seen in an ethanol glass after a nanosecond pulse of fast electrons. The spectrum immediately after the pulse recalls that shown in Fig. 8.1 for electrons trapped in a hydrocarbon glass. The change of shape taking place within a few microseconds is probably caused by relaxation of ethanol molecules.

in the first instance the electrons are mainly trapped in shallow traps, and after the alcohol molecules have reacted to the presence of the charge the centre becomes more like a solvated electron as seen in the solid phase in alkaline ice (p. 112). The relaxation process is much faster than might be expected for 77°K, perhaps because of local heating produced when the electrons slow down.

Optical absorption spectra like the one in Fig. 8.4 with a peak near 550 nm are also seen after the γ-irradiation of glasses made from methanol. ethanol, *n*-propanol and so on at 77°K, and the ESR data and the effect of scavengers confirm that trapped electrons are responsible. The yields of the trapped electrons are usually found to be in the region $G = 1.5$–2.5, and extinction coefficients at the maximum are about 10^4 M^{-1} cm^{-1} [52]. The associated positive ions are probably solvated protons, formed in a reaction of the parent cation analogous to that which forms H_3O^+ in water.

Trapped electrons in methanol, ethanol or *n*-propanol disappear thermally in the dark at, for example, 100°K, without forming detectable additional free radicals. The process may consist of a reaction with the solvent itself. For example in an *n*-propanol glass:

$$e^-_{solv} + C_2H_5CH_2OH \longrightarrow C_2H_5CH_2O^- + H \quad (8.27)$$

Alternatively the electrons may be reacting with solvated protons, also giving H. Since hydrogen atoms are not detected, they may be presumed to have reacted with each other immediately (Reaction 8.7). Trapped electrons may also be photobleached, but in this case hydroxyalkyl radicals are formed in their neutral or anionic forms. This process may consist of reactions like:

$$e^-_{solv} + C_2H_5CH_2OH \longrightarrow C_2H_5\dot{C}HO^- + H_2 \quad (8.28)$$

or

$$e^-_{solv} + 2C_2H_5CH_2OH \longrightarrow C_2H_5\dot{C}HOH + H_2 + C_2H_5CH_2O^- \quad (8.29)$$

where the reactions may or may not involve the intermediate formation of hot hydrogen atoms [53].

When electron scavengers are present in alcohol glasses they can trap the electrons. When anions are formed, as with biphenyl or naphthalene, protonation often follows. This also happens in water, whereas protonation of anions does not appear to take place in hydrocarbon glasses.

In liquid alcohols, free ion pairs are formed in a yield which is estimated from most steady state and pulse radiolysis data to lie in the approximate range 0.5–1.5, although both lower and higher values have been reported. A total yield of ion pairs of about 4.2–4.6 may be assumed by analogy with values of *W* in the vapour phase, and from certain experiments with scavengers like N_2O [54]. Despite the variations, it is evident that, in qualitative agreement with the values of the dielectric constants, the amount of geminate recombination is very much greater than the small amount which must be occurring in water, and very much less than the amount which occurs with most aliphatic hydrocarbons. Aspects of the course of the recombination have been observed by pulse radiolysis of liquid *n*-propanol cooled to 152°K. Under these conditions the absorption spectrum immediately after a 5 ns pulse is very like that shown for glassy ethanol in Fig. 8.4. The whole spectrum changes with a half-life of 60 ns to the one with an absorption maximum near 550 nm, attributable in the main to electrons round which the alcohol molecules have relaxed to form conventional solvated electrons. Solvated electrons then disappear in two stages. The first, with a half-life of 6 μs, is attributable to recombination of some of the electrons with their geminate partners and the second, half-life greater than 1 ms, is due to reaction of free solvated electrons with solvent [55]. The number of solvated electrons disappearing in the geminate process is approximately equal to the number of free solvated electrons. A similar additional number of electrons is thought to recombine geminately before becoming fully solvated. All processes are speeded up at higher temperatures.

The organic free radicals observed in alcohol glasses have uncharacteristic optical absorption spectra rising from about 400 nm towards the ultraviolet. ESR determinations show that the radicals are mainly hydroxyalkyl. The yield is in the region 4–6 [52]. Similar radicals are observable in the liquid by pulse radiolysis. They react with each other at rates which are close to diffusion-controlled.

Products The yields of products detected after irradiation of alcohols are strongly dependent on the purity of the material. The yield of hydrogen from liquid methanol for example can be depressed by as much as 10 per cent by the presence of certain substances at concentrations of only 2 micromolar. Products detected after γ-irradiation of pure methanol at ambient temperature are listed in Table 8.6. The material balance is reasonable providing a small yield of water is assumed to be produced with G about 0.5, although there may also be a small amount of undetected hydrogen-deficient material. The yield of hydrogen is about half that found in the low-pressure vapour phase at the same temperature, and the yields of the other products are also markedly different. Other alcohols also give hydrogen as main product, while aldehydes are formed from primary alcohols and both aldehydes and ketones are formed from secondary alcohols. Glycols, hydrocarbons and so on also appear [54]. The glycols are principally vicinal.

Table 8.6 *Products of the γ-radiolysis of methanol* [56]

Product	*Initial G-value*
Hydrogen	5.39
Carbon monoxide	0.11
Methane	0.54
Formaldehyde	1.84
Ethylene glycol	3.63

The glycols may be assumed to arise from dimerization of hydroxyalkyl radicals:

$$2R\dot{C}HOH \longrightarrow (RCHOH)_2 \tag{8.30}$$

Hydroxyalkyl radicals could be formed from excited or ionized alcohol by simple fission of C—H bonds, as well as by reaction of alkoxyl radicals or, at ambient temperature, by reaction of atomic hydrogen with alcohol:

$$RCH_2O\cdot + RCH_2OH \longrightarrow RCH_2OH + R\dot{C}HOH \tag{8.31}$$

$$H + RCH_2OH \longrightarrow H_2 + R\dot{C}HOH \tag{8.32}$$

The fact that the glycols are vicinal shows that the α C—H bonds are the most easily broken. The aldehydes may arise partly in disproportionation reactions of the organic radicals, and more aldehyde is probably formed through direct unimolecular elimination of H_2 or 2H. Hydrogen itself will be partly produced directly in the molecular form by the unimolecular elimination and partly by reactions like 8.32. Hydrogen atoms come from unimolecular elimination as well as from reaction of solvated electrons with alcohol as in Reaction 8.27 (cf. the forward Reaction 7.33, p. 148). The available quantitative data is not sufficient to enable G-values to be assigned to all the reactions taking place, even for methanol, except on a purely hypothetical basis.

The G-values of all products from methanol are independent of dose in the range 10^{18}–10^{20} eV g^{-1}, but the G-values of hydrogen and acetaldehyde from ethanol and isopropanol decrease in this range. The basic reason is probably that electrons react efficiently with acetaldehyde and propionaldehyde, whereas they do not react efficiently with formaldehyde [54]. The reaction of an electron with an aldehyde not only reduces the amount of aldehyde formed, but also lowers the number of electrons giving rise to H_2 via reactions like 8.27.

The influence of temperature, LET, and added substances such as acid and alkali, have all been studied and attempts have been made to account for the results on the basis of the reactions of the intermediates present [54].

Other organic compounds

As well as the few relatively simple examples discussed here, the radiation chemistry of numerous other organic compounds including halides, cyclic ethers, carboxylic acids and many, many others has been studied. Like the irradiations discussed in other parts of this book, the results give increasing information about the properties and reactions of the excited states, trapped electrons, ions and free radicals produced. The information adds significantly to knowledge of the chemistry of short-lived species which may be gained from other fields such as photochemistry, chemical kinetics and so on.

An exhaustive review of product formation in the irradiation of organic compounds was given in 1960 [57]. Since that time, improved methods of analysis, including especially gas chromatography, have been used to determine the products in greater detail. Also the effect of irradiation conditions has been studied more thoroughly. Some advance has been made in understanding mechanisms. In addition, reasonably full investigations have been made of those effects of radiation on organic compounds which are of possible importance for the nuclear or other industries, to the extent that there is now a reasonable basis for making technological decisions and for planning further work in desired directions.

PROBLEMS

1. Cyclohexane, in a cell of optical path length 4 cm, is given a single sub-microsecond pulse of fast electrons, dose 1.5×10^{17} eV cm^{-3}. A graph based on an oscilloscope trace shows that after the pulse the inverse of the absorbance at 240 nm is equal to 0.44 multiplied by the time after the pulse in microseconds. The intercept at zero time corresponds to an absorbance of 0.088. If the absorbance is attributable solely to cyclohexyl radicals formed with a yield of $G = 5.2$, calculate (*a*) the extinction coefficient of $C_6H_{11}\cdot$ at 240 nm (*b*) the rate constant, k, for disappearance of the radicals, defined by $-d[C_6H_{11}\cdot]/dt = 2k[C_6H_{11}\cdot]^2$ (*c*) the concentration of radicals which would be expected to be present if cyclohexane were to be irradiated in the cavity of an ESR spectrometer at a dose-rate of 5×10^{17} eV cm^{-3} s^{-1}.

[Ans: (*a*) 1700M^{-1} cm^{-1} (*b*) 1.5×10^9M^{-1} s^{-1} (*c*) 1.2×10^{-7}M]

2. What would be the initial rate of production of hydrogen in cm^3 at standard temperature and pressure per kilorad in the γ-irradiation of 10 cm^3 of pure cyclohexane (density 0.78 g cm^{-3})? (*b*) How many cm^3 would be produced by a dose of 3 Mrads?

[Ans: (*a*) 1.0×10^{-3} (*b*) 2.8]

3. Suggest a mechanism to account for the formation of cyclohexyl sulphonic acid ($C_6H_{11}SO_3H$) in the γ-irradiation of cyclohexane through which SO_2 and O_2 are passed.

[Ans: see Reference 26]

4. A determination of the yield of isomerization of *trans*-stilbene in benzene as a function of solute concentration gave the following values.

Stilbene (M)	*G-value*
0.005	1.25
0.0066	1.47
0.01	1.76
0.014	1.93
0.022	2.33
0.05	2.64
0.1	2.78

Make the same assumptions as in the discussion on pp. 188–189 to obtain an independent estimate of the total yield of triplets in the irradiation of pure benzene.

[Ans: $G = 4.8$]

5. It is desired to prepare 10 g of 2:3 dihydroxybutane by irradiating ethanol using a 50 μA beam of 2 MeV electrons. For how long would it be necessary to irradiate if all the energy could be absorbed in the ethanol, and the yield of the product could be taken to be $G = 2.0$?

[Ans: 89 minutes]

REFERENCES

1. F. Hirayama and S. Lipsky. *J. Chem. Phys.*, **51**, 3616–7 (1969), 'Fluorescence of saturated hydrocarbons'.
2. J. E. Willard, in *Fundamental Processes in Radiation Chemistry*, ed. P. Ausloos. Interscience, New York, N.Y., 1968, pp. 599–649, 'Organic compounds in the solid state'.
3. J. B. Gallivan and W. H. Hamill. *J. Chem. Phys.*, **44**, 1279–87 (1966), 'Ionic processes in γ-irradiated organic solids: electrons in 3-methyl pentane'.

4. W. H. Hamill, in *Radical Ions*, eds E. T. Kaiser and L. Kevan. Interscience, New York, N.Y., 1968, pp. 321–416, 'Ionic processes in γ-irradiated organic solids at $-196°$'.
5. B. Brocklehurst and R. D. Russell. *Trans. Faraday Soc.*, **65**, 2159–67 (1969), 'Ion recombination in γ-irradiated naphthalene solutions'.
6. C. Deniau, A. Déroulède, F. Kieffer and J. Rigaut. *J. Luminescence*, **3**, 325–50 (1971), 'Deferred luminescence spectra of some aromatic compounds in rigid organic media'.
7. G. Scholes and M. Simic. *Nature*, **202**, 895–6 (1964), 'Reactions of electrons in the γ-radiolysis of liquid alkanes'.
8. F. Williams. *J. Am. Chem. Soc.*, **86**, 3954–7 (1964), 'Kinetics of ionic processes in the radiolysis of liquid cyclohexane'.
9. J. K. Thomas, K. Johnson, T. Klippert and R. Lowers. *J. Chem. Phys.*, **48**, 1608–12 (1968), 'Nanosecond pulse radiolysis studies of the reaction of ions in cyclohexane solutions'.
10. E. J. Land and A. J. Swallow. *Trans. Faraday Soc.*, **64**, 1247–55 (1968), 'Formation of excited states in the pulse radiolysis of solutions of aromatic compounds in cyclohexane and benzene'.
11. J. L. Magee, in *Comparative Effects of Radiation*, eds M. Burton, J. S. Kirby-Smith and J. L. Magee. Wiley, New York, N.Y., 1960, pp. 130–46, 'Elementary processes in action of ionizing radiation'.
12. J. M. Warman, K.-D. Asmus and R. H. Schuler. *J. Phys. Chem.*, **73**, 931–9 (1969), 'Electron scavenging in the radiolysis of cyclohexane solutions of alkyl halides'.
13. R. W. Fessenden and R. H. Schuler. *Advan. Radiation Chemistry*, **2**, 1–176 (1970), 'Electron spin resonance spectra of radiation-produced radicals'.
14. R. A. Holroyd, in *Aspects of Hydrocarbon Radiolysis*, eds T. Gäumann and J. Hoigné. Academic Press, New York, N.Y., 1968, pp. 1–32, 'Radical yields of hydrocarbons'.
15. S. K. Ho and G. R. Freeman. *J. Phys. Chem.*, **68**, 2189–97 (1964) 'Radiolysis of cyclohexane. V. Purified liquid cyclohexane and solutions of additives'.
16. K. H. Jones. *J. Phys. Chem.*, **71**, 709–11 (1967), 'Radiolysis of cyclohexane and a mixture of cyclohexane and benzene'.
17. R. W. Fessenden and R. H. Schuler. *J. Chem. Phys.*, **39**, 2147–95 (1963), 'Electron spin resonance studies of transient alkyl radicals'.
18. R. A. Holroyd, in *Fundamental Processes in Radiation Chemistry*, ed. P. Ausloos. Interscience, New York, N.Y., 1968, pp. 413–514, 'Organic liquids'.
19. P. J. Dyne and W. M. Jenkinson. *Can. J. Chem.*, **38**, 539–43 (1960), 'Radiation chemistry of cyclohexane. I. Isotopic composition of hydrogen evolved from mixtures of C_6D_{12} and C_6H_{12}'.
20. R. H. Schuler. *J. Phys. Chem.*, **61**, 1472–6 (1957), 'The effect of solutes on the radiolysis of cyclohexane'.
21. W. G. Burns and C. R. V. Reed. *Trans. Faraday Soc.*, **66**, 2159–81 (1970), 'Effects of L.E.T. and temperature in the radiolysis of cyclohexane, Part I. Experimental results and comparison with predictions of diffusion kinetic models'.
22. H. A. Dewhurst. *J. Am. Chem. Soc.*, **83**, 1050–2 (1961), 'Radiolysis of organic compounds, V. *n*-Hexane vapor'.
23. H. Widmer and T. Gäumann. *Helv. Chim. Acta*, **46**, 2766–80 (1963), 'Strahlungschemie der Kohlenwasserstoffe 7. Jod in Hexan'.

24. J. H. Futrell. *J. Am. Chem. Soc.*, **81**, 5921–4 (1959), 'High energy electron irradiation of *n*-hexane'.
25. C. D. Wagner. *Advan. Radiation Chemistry*, **1**, 199–244 (1969), 'Chemical synthesis by ionizing radiation'.
26. D. O. Hummel, W. Mentzel and C. Schneider. *Liebigs Ann. Chem.*, **673**, 13–26 (1964), 'Über die strahlenchemische Sulfoxydation von Cyclohexan'.
27. J. P. Guarino and W. H. Hamill. *J. Am. Chem. Soc.*, **86**, 777–81 (1964), 'Ionic intermediates in γ-irradiated organic glasses at $-196°$'.
28. D. R. Smith and J. J. Pieroni. *J. Phys. Chem.*, **70**, 2379–83 (1966), 'Observations on trapped electrons and allyl radicals formed in 2-methylpentene-1 by γ-radiolysis at low temperature'.
29. B. R. Wakeford and G. R. Freeman. *J. Phys. Chem.*, **68**, 2635–8 (1964), 'Radiolysis of cyclohexene. I. Pure liquid'.
30. P. C. Chang, N. C. Yang and C. D. Wagner. *J. Am. Chem. Soc.*, **81**, 2060–4 (1959), 'Direct dimerization of terminal olefins by ionizing radiation'.
31. F. Williams, in *Fundamental Processes in Radiation Chemistry*, ed. P. Ausloos. Interscience, New York, N.Y., 1968, pp. 515–98, 'Principles of radiation-induced polymerization'.
32. W. G. Burns and J. A. Winter. *Discussions Faraday Soc.*, **36**, 124–34 (1963), 'LET dependence of G values as a guide in determining the primary mechanisms of radiolysis. L.E.T. effects in cyclohexene'.
33. M. Brun and R. Montarnal. *Compt. rend.*, **247**, 2361–4 (1958), 'Etude cinétique de l'oxydation radiochimique du cyclohexène en phase liquide'.
34. A. T. Betts and N. Uri. *Nature*, **199**, 568–9 (1963), 'Some unusual observations in a comparison of liquid and solid phase autoxidation'.
35. R. Voltz. *Radiat. Res. Rev.*, **1**, 301–60 (1968), 'Electronic energy transfer in irradiated aromatic materials'.
36. G. K. Oster and H. Kallmann. *J. chim. phys.*, **64**, 28–32 (1967), 'Energy transfer from high-lying excited states'.
37. R. Cooper and J. K. Thomas. *J. Chem. Phys.*, **48**, 5097–102 (1968), 'Formation of excited states in the nanosecond-pulse radiolysis of solutions of benzene and toluene'.
38. M. A. Dillon and M. Burton, in *Pulse Radiolysis*, eds M. Ebert, J. P. Keene, A. J. Swallow and J. H. Baxendale. Academic Press, New York, N.Y., 1965, pp. 259–77, 'Excitation transfer and decay processes in multi-component systems: cyclohexane + benzene + *p*-terphenyl'.
39. P. Skarstad, R. Ma and S. Lipsky. *Mol. Cryst.*, **4**, 3–14 (1968), 'The scintillation efficiency of benzene'.
40. R. R. Hentz, D. B. Peterson, S. B. Srivastava, H. F. Barzynski and M. Burton. *J. Phys. Chem.*, **70**, 2362–72 (1966), 'The radiation-induced isomerization of stilbene in benzene and cyclohexane'.
41. R. R. Hentz and L. M. Perkey. *J. Phys. Chem.*, **74**, 3047–54 (1970), 'Yields of the lowest triplet and excited singlet states in γ radiolysis of liquid benzene'.
42. S. Arai, D. A. Grev and L. M. Dorfman. *J. Chem. Phys.*, **46**, 2572–8 (1967), 'Pulse radiolysis studies. X. Electron transfer reactions of aromatic molecules in solution'.
43. R. L. McCarthy and A. MacLachlan. *Trans. Faraday Soc.*, **56**, 1187–200 (1960), 'Transient benzyl radical reactions produced by high-energy radiation'.
44. V. V. Voevodskii and Yu. N. Molin. *Radiat. Res.*, **17**, 366–78 (1962), 'On the radiation stability of solid organic compounds'.

45. M. Burton and W. N. Patrick. *J. Chem. Phys.*, **22**, 1150 (1954), 'Radiation chemistry of luminescent solutions'.
46. J. Hoigné, in *Aspects of Hydrocarbon Radiolysis*, eds T. Gäumann and J. Hoigné. Academic Press, New York, N.Y., 1968, pp. 61–151, 'Aromatic hydrocarbons'.
47. W. G. Burns. *Trans. Faraday Soc.*, **58**, 961–70 (1962), 'Decomposition of aromatic substances by different kinds of radiation. Part 1. Proton irradiation of benzene'.
48. R. H. Schuler. *Trans. Faraday Soc.*, **61**, 100–9 (1965), 'Radiolysis of benzene by heavy ions'.
49. J. G. Carroll, R. O. Bolt and C. A. Trilling, in *Radiation Effects on Organic Materials*, eds R. O. Bolt and J. G. Carroll. Academic Press, New York, N.Y., 1963, pp. 289–348, 'Coolants'.
50. J. G. Carroll and R. O. Bolt, in *Radiation Effects on Organic Materials*, eds R. O. Bolt and J. G. Carroll. Academic Press, New York, N.Y., 1963, pp. 349–406, 'Lubricants'.
51. J. T. Richards, and J. K. Thomas. *J. Chem. Phys.*, **53**, 218–24 (1970), 'Trapping of electrons in low-temperature glasses. A pulse radiolysis study'.
52. L. Kevan, in *Actions Chimiques et Biologiques des Radiations*, treizième série, ed. M. Haïssinsky. Masson, Paris, 1969, pp. 57–117, 'Radiation chemistry of frozen polar systems'.
54. F. S. Dainton, G. A. Salmon and U. F. Zucker. *Proc. Roy. Soc. (London)*, **A 320**, 1–22 (1970), 'The radiation chemistry of glassy *n*-propanol'.
54. G. R. Freeman, in *Actions Chimiques et Biologiques des Radiations*, quatorzième série, ed. M. Haïssinsky. Masson, Paris, 1970, pp. 73–134, 'The radiolysis of alcohols'.
55. J. H. Baxendale and P. Wardman. *Nature*, **230**, 449–50 (1971), 'Direct observation of solvation of the electron in liquid alcohols by pulse radiolysis'.
56. G. Meshitsuka and M. Burton. *Radiat. Res.*, **8**, 285–97 (1958), 'Radiolysis of liquid methanol by Co^{60} gamma-radiation'.
57. A. J. Swallow. *Radiation chemistry of Organic Compounds*. Pergamon Press, New York, N.Y., 1960.

Polymeric systems

Irradiation of monomers and polymers causes excitation and ionization leading to chemical change, and these effects are in many ways similar to those occurring with other substances. What is most interesting however, is that the molecular weight alters. The monomers polymerize, and the molecules of the polymers link together or break up, in all cases leading to striking changes in properties. These effects are of some industrial interest. Furthermore the results of irradiating relatively simple polymeric materials may provide clues as to the radiation response of biological macromolecules. For these reasons a very great deal of research has been done in this field. The work has been fruitful, and the radiation chemistry of polymeric systems was one of the first branches of radiation chemistry to reach comparative maturity [1–3].

Polymerization

Between 1938 and 1940, styrene, methyl methacrylate and vinyl acetate were irradiated for the first time, using γ-rays and fast neutrons, and were found to give high polymers [4]. Now excited molecules may give rise to radicals, while the parent positive ions formed from unsaturated compounds may be regarded as free radicals at one end of the molecule and carbonium ions at the other:

$$\overset{+}{\diagdown}\!\!\!\!\!\!\;\; +\mathrm{C}{-}\mathrm{C}\cdot$$

Similarly, any anion formed by electron capture by an unsaturated compound is a free radical as well as an anion. The chain polymerization of vinyl monomers by conventional means can take place either by free-radical or by ionic mechanisms, so in principle either mechanism might operate under irradiation. Until about 1956–57 the experimental evidence pointed to all the polymerizations induced by radiation being free-radical reactions, but since about 1957 it has become apparent that ionic polymerizations can also take place, providing the conditions are correctly chosen.

Reaction mechanism There are several pieces of evidence concerning the reaction mechanisms. First, the work of numerous investigators has shown that compounds such as oxygen and benzoquinone, which are known to be inhibitors of free-radical reactions, often inhibit the polymerizations produced by radiation. Such an inhibition has been noted for the room temperature polymerization of conventionally purified styrene, methyl methacrylate, vinyl acetate, acrylonitrile, vinyl chloride and many other monomers. Benzoquinone also inhibits the polymerization of styrene irradiated in solution in solvents such as methanol, carbon tetrachloride and nitrobenzene [3]. These reactions could therefore be free-radical reactions although the evidence is not in itself sufficient to constitute proof. Additional evidence is provided by the temperature dependence of the rate of reaction. Most monomers polymerize more quickly under irradiation as the temperature increases, with activation energies the same as those for the known free-radical reaction. This has been shown for styrene and methyl methacrylate [5], and has also been found for 2,4-dimethylstyrene, acrylonitrile, methacrylonitrile, formaldehyde, methyl acrylate, vinyl acetate, vinyl chloride, vinyl stearate, vinylcarbazole, vinylpyrrolidone and many other monomers in the liquid phase. The activation energies are in the region of 5 kcal per mole.

Early evidence for an ionic mechanism was obtained in the radiation-induced polymerization of styrene dissolved in methylene chloride at −78°C [6]. Air did not inhibit the polymerization but substances such as methanol and acetone, which are known to inhibit cationic reactions, did act as inhibitors. At such low temperatures, free-radical reactions are very much reduced in rate whereas ionic mechanisms are not, so it seems as if ionic mechanisms are able to take over under these conditions. Rigorous drying is another factor which can greatly favour ionic mechanisms [7]. Some experiments have shown that additives such as silica gel and zinc oxide can assist ionic polymerization to take place, and this is probably, in part at least, because of their capacity to take water out of the system.

Copolymerization studies provide further evidence concerning the reaction mechanism. The relative rates of reaction of a growing polymer chain towards two monomers is known to depend on whether the chain has been initiated by a free-radical, an anionic or a cationic catalyst. In the particular case of an equimolar mixture of styrene and methyl methacrylate, free-radical initiation leads to a copolymer containing 50 per cent of each component, anionic initiation to a copolymer containing 99 per cent methyl methacrylate and cationic initiation to a copolymer containing 99 per cent styrene. The irradiation of an equimolar mixture of conventionally purified styrene and methyl methacrylate at room temperature yields a 50 per cent copolymer, in agreement with a free-radical mechanism [8]. On the other hand this technique provides evidence for ionic mechanisms under conditions which depress the reactivity of free radicals and favour that of ions, for example a copolymer containing more than 99 per cent styrene was formed in the irradiation of an equimolar mixture of styrene and methyl methacrylate dissolved in methylene chloride at −78°C [6].

One of the most important factors determining the reaction mechanism is the nature of the monomer. Some monomers such as styrene are able to polymerize by either radical or ionic mechanisms, but certain cyclic substances including epoxides, ethers, imines and sulphides are thought to polymerize only by ionic mechanisms. The polymerization of hexamethylcyclotrisiloxane by fast electrons was an early indication that high-energy radiation could initiate ionic chain reaction [9]. Similarly isobutylene, which is known to polymerize by a cationic mechanism at low temperatures, is found to polymerize readily under γ- or electron irradiation at −80°C [10]. Nitroethylene provides an example of a substance polymerizing by an anionic mechanism under irradiation [11].

Kinetics of free radical polymerization in the liquid phase It is clear that in many cases, radiation-induced polymerizations take place predominantly by free-radical mechanisms. Considerable success has been achieved in applying the kinetics of free-radical chain reactions to these processes [3, 12], thus providing still further evidence that in appropriate cases the reactions are free-radical ones.

There are three basic steps in the polymerization. The first is the formation of initiating radicals by the radiation. With low LET radiation, as used in the vast majority of polymerization studies, the radicals may be assumed to be dispersed essentially uniformly throughout the liquid. The radicals are formed both from the monomer itself and from the solvent if any is present:

$$M, S \rightsquigarrow R\cdot \qquad (9.1)$$

The production of radicals depends on the *G*-values for radical formation from monomer and solvent but, owing to those effects which may loosely be referred to as 'energy transfer', is rarely an additive function of composition. Reaction 9.1 is the only initiation step at the beginning of the polymerization but at high conversions the radiation gives radicals by direct action on the polymer itself.

The next step is chain propagation. It consists of the addition of the initiating radical to the double bond of the monomer, giving another radical which in turn adds to a monomer molecule giving another radical, and so on:

$$R\cdot + M \longrightarrow RM\cdot \qquad (9.2)$$

$$RM\cdot + M \longrightarrow RM_2\cdot \qquad (9.3)$$

In general:

$$RM_n\cdot + M \longrightarrow RM_{n+1}\cdot \qquad (9.4)$$

The final step is chain termination. In homogeneous liquid phase this occurs by the mutual interaction of two growing polymer radicals either by combination (dimerization) or by disproportionation, in each case giving 'dead' polymer:

$$RM_n\cdot + RM_m\cdot \longrightarrow P \qquad (9.5)$$

Under stationary state conditions the rate of formation of initiating radicals, v_i, is equal to the rate of removal of radicals by termination:

$$v_i = k_t[RM_n\cdot]^2 \qquad (9.6)$$

where k_t is the rate constant for Reaction 9.5 as defined by $-d[RM_n\cdot]/dt = k_t[RM_n\cdot]^2$, and n may have any value. Providing the chain length is long, the consumption of monomer is governed almost exclusively by the propagation step. This rate, v_p, which is also equal to the rate of formation of polymer, is thus given by:

$$v_p = k_p[RM_n\cdot][M] \qquad (9.7)$$

where k_p is the rate constant for Reaction 9.4.

Combining Equations 9.6 and 9.7, we obtain:

$$v_p = k_p k_t^{-0.5} v_i^{0.5}[M] \qquad (9.8)$$

The rate of initiation, v_i, is equal to the rate at which free radicals are formed in the system and is therefore proportional to the *G*-value for radical formation in the system and to the dose-rate. Hence v_p, the overall rate of polymerization, is proportional to the square root of the dose-rate. Such a relationship has been found experimentally in numerous cases where a monomer is polymerized in the homogeneous liquid phase.

Typical results, for the polymerization of conventionally purified styrene at 19°C, are shown in Fig. 9.1 from which it can be seen that, in accordance with the theory, log v_p/log dose-rate is close to 0.5 over a wide range of dose rates [13].

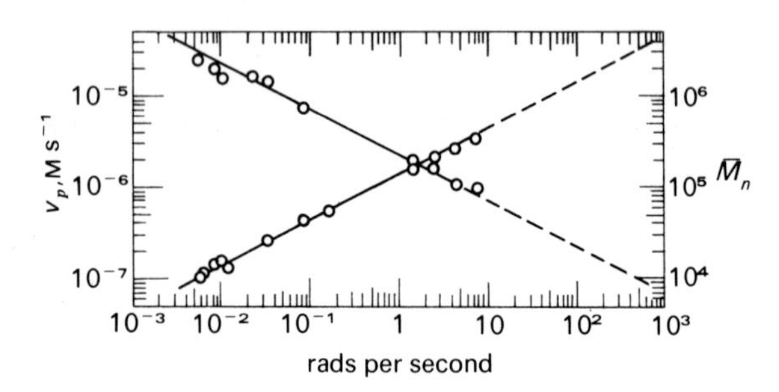

Fig. 9.1 Dependence on dose-rate of the rate of polymerization and the molecular weight of polymer in the free-radical polymerization of styrene.

The kinetic scheme can also be applied to the molecular weight of the polymer. If Reactions 9.1–9.5 are the only ones occurring, then the kinetic chain length is identical with the number-average degree of polymerization, $\overline{P}_n$.† The chain length is equal to the amount of polymer formed divided by the number of chains initiated. If termination is by disproportionation, then:

$$\overline{P}_n = v_p/v_i = k_p k_t^{-0.5} v_i^{-0.5}[M] \tag{9.9}$$

Multiplication of $\overline{P}_n$ by the molecular weight of the monomer gives the number-average molecular weight of the polymer, $\overline{M}_n$. If termination is by combination instead of by disproportionation, $\overline{P}_n$ becomes twice the right hand side of Equation 9.9. The dependence of $\overline{M}_n$ on dose-rate, as shown for styrene in Fig. 9.1, is often found to conform to the simple theory.

It is possible for chains to terminate by reaction with monomer, added substances or, at higher doses, already formed polymer, giving another radical. This process is known as chain transfer. If this process occurs:

$$RM_n\cdot + AB \longrightarrow RM_nA + B\cdot \tag{9.10}$$

† The degree of polymerization is the number of repeating units in a chain, that is, the molecular weight divided by the molecular weight of the repeating unit. Number-average degree of polymerization is $\sum n_i P_i / \sum n_i$ where n_i is the number of molecules of degree of polymerization P_i. Weight-average degree of polymerization is $\sum n_i P_i^2 / \sum n_i P_i$.

the rate of polymerization remains the same as in the absence of chain transfer providing the radical B can initiate another chain,† but the degree of polymerization decreases, being given for chain transfer to monomer and added substances by:

$$\frac{1}{\bar{P}_n} = \frac{k_t^{0\cdot5}}{k_p}\frac{v_i^{0\cdot5}}{[M]} + \frac{k_{trm}}{k_p} + \frac{k_{trs}}{k_p}\frac{[S]}{[M]} \qquad (9.11)$$

where k_{trm} and k_{trs} are rate constant for transfer to monomer and the added substance respectively, and $[S]$ is the concentration of the added substance. Very efficient chain transfer to an added substance decreases the molecular weight so much that the reaction product can hardly be called a polymer, and is usually called a 'telomer'. In the extreme case the chain transfer Reaction 9.10 almost always takes place in preference to Reaction 9.2. The reaction then becomes a 1:1 addition to the unsaturated double bond. This reaction has already been mentioned in Chapters 6 (p. 132) and 8 (p. 185). A further example is the chain addition of n-butyl mercaptan to 1-pentene [14]:

$$C_3H_7\dot{C}H{-}CH_2SC_4H_9 + C_4H_9SH \longrightarrow C_3H_7CH_2{-}CH_2SC_4H_9 + C_4H_9S\cdot \qquad (9.12)$$

followed by:

$$C_4H_9S\cdot + C_3H_7CH{=}CH_2 \longrightarrow C_3H_7\dot{C}H{-}CH_2SC_4H_9 \qquad (9.13)$$

and so on, giving the overall reaction:

$$C_4H_9SH + C_3H_7CH{=}CH_2 \longrightarrow C_3H_7CH_2{-}CH_2SC_4H_9 \qquad (9.14)$$

Careful note should be made of the assumptions implied in deriving the kinetic equations given in this section. Where the assumptions do not apply, as at very high dose-rates, the equations must eventually break down [3, 12].

The gel effect At very high conversions, the viscosity of a polymerizing system becomes very high. This hinders termination by mutual interaction of growing chains (Reaction 9.5) but has less effect on the propagation reaction 9.4 because diffusion of the small monomer molecules is not so much affected by viscosity. Hence both the rate of polymerization and the molecular weight of the polymer show an increase at high conversions. This effect is known as the gel effect or Trommsdorff effect. A similar effect is evident whenever a growing polymer precipitates out of the medium in which it was formed. Termination then becomes highly hindered at the expense of propagation, so that the rate of polymerization

† If it can not, the reaction is called 'degradative chain transfer'.

increases above that expected on the square-root law. In the case of acrylonitrile in water, for example, the rate of polymerization depends on a power of the dose-rate which can be as high as about 0.85 at low dose-rates [15]. The radicals which precipitate out of the system can be extremely long lived in certain cases, and polymerization can continue for as much as several days after irradiation has stopped.

Kinetics of ionic polymerization in the liquid phase [12]. Consider the formation of ions by the action of radiation on the monomer:

$$M \rightsquigarrow M^+ + e^- \quad (9.15)$$

Suppose the electron reacts to form a non-propagating species, while the positive ions can initiate polymerization by transferring a proton to monomer or by adding on to another monomer molecule:

$$M^+ + M \longrightarrow M_2^+ \quad (9.16)$$

leading to the general propagation reaction:

$$M_m^+ + M \longrightarrow M_{m+1}^+ \quad (9.17)$$

Suppose furthermore that an impurity, X, is present which is capable of terminating the chain:

$$M_m^+ + X \longrightarrow M_m + X^+ \quad (9.18)$$

The termination may occur by charge transfer or more likely proton transfer or addition of an ion to a base. By considerations similar to those discussed for free-radical polymerization we can obtain:

$$v_p = k_{pc} k_{tx}^{-1} [X]^{-1} v [M] \quad (9.19)$$

where k_{pc} and k_{tx} are the rate constants for Reactions 9.17 and 9.18 respectively. This expression shows the rate of polymerization, v_p, to be proportional to the first power of v_i, that is, to the first power of dose-rate. An analogous expression would hold for anionic polymerization. Impurity termination in free-radical polymerization would also lead to a similar expression, but in free-radical polymerization any impurities are generally consumed at the beginning of the irradiation so that after a brief period of inhibition, Equation 9.8 may hold and v_p would tend to become proportional to the square root of the dose-rate. However, in ionic polymerization the impurity X (which may be a trace of water for example) appears to be capable of being regenerated in the charge neutralization process so that except at extreme degrees of purity, the kinetics are expected to obey Equation 9.19.

If the amount of impurity is at a very low level it becomes necessary to consider termination by other processes. In conventional ionic polymerization, the propagating ion and the counter ion (gegenion) keep together when the dielectric constant of the medium is low and the solvating power is weak, whereas they separate when the dielectric constant is high and the solvating power is strong, and the polymerization becomes produced by free ions. It is the latter process which appears to take place under irradiation, although ion pair propagation has been considered. If the ions are regarded as separate, termination would occur by the general reaction:

$$M_m^{\ +} + M_n^{\ -} \longrightarrow P \qquad (9.20)$$

and for the case of cationic polymerization, and neglecting Reaction 9.18, we would obtain:

$$v_p = k_{pc} k^{-0.5} v_i^{0.5} [M] \qquad (9.21)$$

where k_t is the rate constant for Reaction 9.20, that is, v_p would be proportional to the square root of the dose-rate. An analogous expression would hold for anionic polymerization. Fig. 9.2 shows the expected dose-rate dependence for the cationic polymerization of styrene, calculated for termination by Reactions 9.20 and 9.18 at various impurity concentrations. The rate constants assumed are based on experimental values [16].

An important characteristic of ionic polymerization under irradiation is that the propagation rate constants are several orders of magnitude more than for free-radical polymerization or polymerization by ion pairs. Also the activation energies are very low. Consequently, providing the concentration of ionic terminator is kept down (for example, by extremely rigorous drying of monomer and reaction vessel) the observed rates of polymerization may be perhaps thousands of times greater (depending on temperature and dose-rate) than for any free-radical polymerization which might take place in the presence of ordinary laboratory concentrations of impurity.

Polymerization in the solid state [17] Numerous monomers have been found to polymerize when irradiated in the solid state. One of the first examples was acrylamide, melting point 84°C, for which radiation-induced polymerization within the crystal was demonstrated by X-ray diffraction analysis as well as by conventional polymer techniques [18]. Hexamethylcyclotrisiloxane, melting point 64°C, provides another example [9]. Among the other monomers which have been polymerized by irradiation in the solid state, sometimes at temperatures as low as −196°C, are styrene, vinyl acetate, isoprene and acrylonitrile, as well as substances like formaldehyde, acetaldehyde, acetone, trioxan and acetonitrile. It has been known since the nineteen thirties that polymerization can be induced in

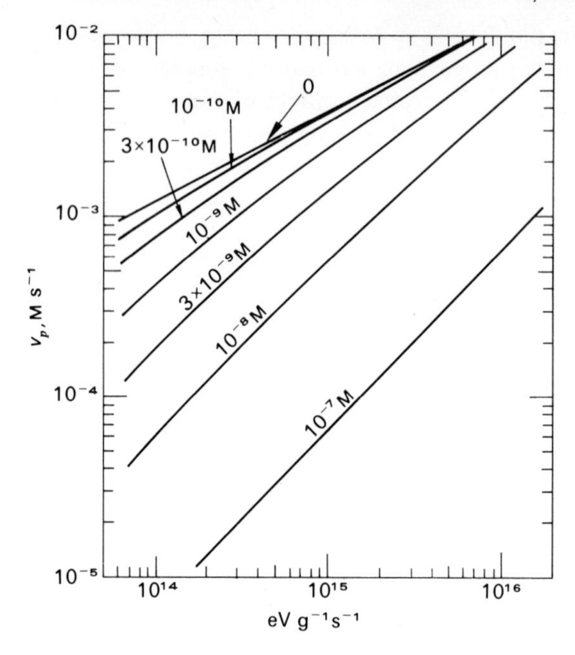

Fig. 9.2 Dependence on dose-rate of the rate of cationic polymerization of styrene at various concentrations of impurity. The curves are calculated assuming $k_t = 2\times10^{11}$ M^{-1} s^{-1}, $k_{tx} = 10^{10}$ M^{-1} s^{-1}, $k_{pc} = 5\times10^{6}$ M^{-1} s^{-1} and G(ion-pairs) = 0.1.

the solid state by chemical means. Thermal, mechanical, catalytic and photochemical methods have also been used. But irradiation is a particularly convenient way of inducing polymerization in a solid.

At first sight it is perhaps surprising that polymerization should be able to occur so readily in a solid. In some cases the orientation of the monomer molecules will be such that only a small movement will enable them to link together to form the polymer chain. This appears to be the case with acetaldehyde [19]. But in most cases it is necessary for the molecules to move about a nm if they are to join a growing chain, and this process might be expected to be strongly hindered in the solid phase. It has indeed been shown that polymerization can occur after irradiation, consequent on the melting or dissolving which is needed to enable certain measurements to be made on the polymer. However, the use of physical techniques which can be applied directly to the irradiated solids, like X-ray diffraction, electron microscopy and optical birefringence proves that polymerization can occur in the solid phase itself. One factor is that the heat of polymerization might be sufficient to permit local melting, but polymerization occurs on room-temperature irradiation of potassium acrylate, melting point 360°C, for which it has been shown that the heat of polymerization of a growing chain could not 'melt' more than one molecule [20]. Consequently the polymerization has to be regarded as a genuine phenomenon of the solid state.

One of the most important factors in solid state polymerization is the precise structure of the solid. Solids which can exist in both the glassy and the crystalline state show marked differences according to the phase. Because of this, samples irradiated under identical conditions can show differences in polymerization behaviour according to the degree of supercooling present. In crystalline solids, defects play an important role, either increasing the rate of polymerization by favouring the mobility of monomer molecules, or decreasing the rate by terminating the chain. The edges of the crystals may also terminate the chain so that crystal size is important. Numerous effects attributable to these factors have been observed, with a variety of monomers, in solid state polymerization induced by radiation as well as by other means [21].

It is very difficult to establish the mechanism of radiation-induced polymerization in the solid phase, since many of the diagnostic methods which have been used in the liquid phase cannot be used unambiguously in the solid phase. From the evidence available at present it seems as if both free-radical and ionic polymerization can occur, depending on the monomer and on the conditions of irradiation. Regardless of the mechanism, one of the most interesting features of solid state polymerization is the influence exerted by the structure of the solid on the polymer formed. When little change in the distance between the units takes place on polymerization, the structure of a monomer lattice can impose order and orientation on the polymer. Such an effect has been noted for example with certain ring compounds [22]. When the distance between units is smaller in the polymer than in an original monomer crystal, so that monomer molecules have to move to become incorporated in the polymer, a volume contraction takes place on polymerization and the polymers produced are amorphous [17].

Polymerization under complex conditions Radiation is a highly versatile initiator of polymerization and, in addition to the relatively simple examples discussed above, has also been used to initiate polymerization in emulsion, in the vapour phase (cf. ethylene, p. 131) and other states. One of the most intensively studied cases is graft polymerization, referred to below (p. 225). Stereospecific polymers can be obtained from several monomers, of which one is 2,3-dimethylbuta-1,3-diene, by irradiating complexes of monomer with urea or thiourea. In these complexes the monomers are stacked up in an orderly manner along canals in the solid. Monomers do not always polymerize under these conditions even when complexes are formed, a limitation which is probably inherent in such a template polymerization [23, 24].

Polymers

Although scattered observations had been made earlier, little work was done on the irradiation of polymers until after the Second World War, when the continued development of nuclear reactors made it necessary to test polymeric materials for stability under irradiation. It soon became clear that although the effects of radiation on polymers were sometimes harmful, they were very often beneficial and, since radiation sources were by now quite plentiful, this opened up the possibility of commercial exploitation. There was still more interest in the field because the reactions occurring in synthetic polymers must be somewhat analogous to those which occur when macromolecules of biological interest are irradiated. The results obtained were found to be of great scientific interest, quite apart from any applications, and the radiation chemistry of polymers now forms an important branch both of radiation chemistry and of polymer science.

Chemical changes in irradiated polymers Chemical changes are central to an understanding of the effects of radiation on polymers. Chemical changes are the basis for the explanation of physical changes, and the establishment of the exact nature of a chemical change is a necessary prerequisite to an understanding of the full reaction mechanism. The changes are generally analogous to those found with the more tractable low molecular weight materials of similar composition. The most important effects, first discussed in detail for polyethylene,† are crosslinking, which is analogous to dimerization, and degradation, which is analogous to main-chain scission [25]. In most polymers one of these processes predominates. If crosslinking predominates, the ultimate effect of

† Polyethylene is a polymer of formula $(CH_2)_n$, made by polymerizing ethylene catalytically at low pressures (high-density polyethylene) or at high pressures (low-density polyethylene). The molecular weight is in the region 10^4–10^6. The main chain consists of $—CH_2—$ units, but some vinylidine ($—CR{=}CH_2$), vinyl ($—CH{=}CH_2$) and vinylene ($—CH{=}CH—$) unsaturation is present, and there is some branching. The material is partly crystalline and partly amorphous. High-density polyethylene is less branched and more crystalline than low-density polyethylene.

irradiation will be to produce a network polymer in which all molecules are joined to each other. If degradation predominates, then the molecules become smaller and smaller as the irradiation proceeds, and the material loses its polymeric properties. The main response for several polymers is shown in Table 9.1. Steric factors must be at least partly responsible for the different behaviour of different polymers. In the case of vinyl polymers

Table 9.1 *Effect of radiation on polymers* [26]

Predominant crosslinking	*Predominant degradation*
Polyethylene	Polyisobutylene
Polypropylene	
Poly(vinyl chloride)	Poly(vinylidine chloride)
Chlorinated polyethylene	Polychlorotrifluoroethylene
Chlorosulphonated polyethylene	Polytetrafluoroethylene
Polyacrylonitrile	Polymethacrylonitrile
Poly(acrylic acid)	Poly(methacrylic acid)
Polyacrylates	Polymethacrylates
Polyacrylamide	
Polyvinylpyrrolidone	
Poly(vinyl alkyl ethers)	
Poly(vinyl methyl ketone)	
Polystyrene	Polyα-methylstyrene
Sulphonated polystyrene	
Natural rubber	Cellulose plastics
Synthetic rubber (except polyisobutylene)	
Polysiloxanes	
Polyamides	
Poly(ethylene oxide)	
Polyesters	

in particular, it can be seen that crosslinking predominates when the formula is $[-CH_2-CHR-]_n$, but degradation predominates when the formula is $[-CH_2-C(CH_3)R-]_n$ [27] and this is probably because the methyl group introduces a steric strain into the molecule, so weakening the C—C links in the main chain and enhancing main-chain scission.

The degree of unsaturation of a polymer can also change on irradiation. Polymers which are initially highly unsaturated (for example, natural rubber) tend to become less unsaturated on irradiation, while with initially saturated polymers the amount of unsaturation increases with dose. The most thorough studies have been made with polyethylene. With this polymer the relatively small amounts of vinyl and vinylidene unsaturation which are present initially are rapidly removed on irradiation, while *trans*-vinylene (main-chain) unsaturation, is produced [28].

The formation of an unsaturated double bond in a polymer will tend to labilize the bonds in the allyl position, so that the formation of further double bonds will tend to occur adjacent to the first, leading to a conjugated system. Such a process accounts for the colouration of some polymers on irradiation. The formation of trapped free radicals can also produce a colour in some cases. Poly(vinyl chloride) becomes particularly

Table 9.2 *Production of gas on irradiation of various materials* [29]

Material	*Molecules of gas per 100 eV*
Polyethylene	3.1
Polystyrene	0.08
Polyα-methylstyrene	0.08
Natural rubber	~0.3
Styrene-butadiene rubber	0.15
Styrene-butadiene plastic	~0.08
Polyisobutylene rubber	~0.8
Nylon	1.1
Aniline-formaldehyde polymer	~0.08
Melamine-formaldehyde polymer (cellulosic filler)	0.45
Urea-formaldehyde polymer (cellulosic filler)	0.8
Nitrile-butadiene rubber	~0.15
Casein plastic	0.15
Poly(methyl methacrylate)	1.5
Poly(ethylene terephthalate)	0.15
Alkyl diglycol carbonate	1.9
Polyesters (general)	0.08–1.9
Cellulose acetate polymer	0.8
Cellulose acetate-butyrate polymer	1.2
Cellulose propionate polymer	1.5
Cellulose nitrate polymer	4.6
Ethyl cellulose polymer	1.5
Phenolic plastic (no filler)	0.1
Phenolic plastic (cellulose filler)	0.8
Phenolic plastic (mineral filler)	<0.08
Silicone elastomer	0.9
Ethyl acrylate rubber	~1.2
Chloroprene rubber	~0.15
Poly(vinyl formal)	~4.3
Triallyl cyanurate polymer	~0.08
Polysulphide rubber	~0.23

coloured on irradiation, becoming yellow, green, red and so on, according to purity. Colour changes in plastics may be employed for dosimetry (p. 53).

Another important chemical change is the formation of gas. The amount of gas produced depends on the nature of the polymer and also on dose, temperature, type of radiation and so on. Typical results for several polymers are shown in Table 9.2. Gas production gives rise to very striking phenomena in the case of poly(methyl methacrylate) (Lucite, Perspex, Plexiglas): irradiation of this material to several Mrads produces little visible change except for the development of a yellow colour, but if the irradiated material is heated above the glass transition temperature† it expands and forms a white brittle material consisting of a mass of bubbles [30] (Fig. 9.3). If the irradiated material is not heated but is left to stand for some months, bubbles do not appear but cracks develop instead.

Fig. 9.3 Appearance of irradiated poly (methyl methacrylate). The sample was originally a small clear block. Irradiation turned it yellow but otherwise there was no change in appearance or dimensions. Heating the irradiated sample caused bubbles of gas to form and expand the material.

Quantitative results of one attempt at measuring all the major changes in polyethylene are shown in Table 9.3. The yields of crosslinks had been found using a modification of the Charlesby-Pinner function (p. 222). The results in the table are for a dose of 27 Mrads: at lower doses the yield of

† The glass transition temperature, or second-order transition point, is the temperature above which micro-Brownian movements occur in an amorphous polymer. Below this temperature the polymer is hard and brittle; above it, it is soft.

crosslinks appears lower and the yield of unsaturation greater. The yield of chain scission is about a quarter of the yield of crosslinks. If a crosslink consists of the formation of a single intermolecular C—C bond and processes other than those in Table 9.3 do not have to be considered, the following material balance equation should hold:

$$G(H_2) = G(\text{crosslinks}) + G(\textit{trans}\text{-vinylene}) + 2G(\text{diene}) \quad (9.22)$$

Examination of the figures in Table 9.3 shows the material balance to be reasonable.

Table 9.3 *Yields of some chemical changes in irradiated polyethylene at 27 Mrads* [31]

Temperature	*G (crosslinks)*	*G (trans-vinylene)*	*G (diene)*	*G (H_2)*
35°	1.5	1.4	0.15	3.3
120°	2.3	1.4	0.15	4.0

Mechanism of the changes in polymers [32] The mechanism by which the chemical changes in polymers are produced must be essentially the same as in molecules of similar chemical composition but of lower molecular weight, but modified by the solid structure. Free radicals must play an important part, and have been extensively studied by means of electron spin resonance. The simplest and most discussed case has been polyethylene, which may be compared with cyclohexane and other saturated hydrocarbons (pp. 167–181). It is highly likely that the crosslinking is due principally to the dimerization of adjacent free radicals:

$$\begin{matrix} -CH_2\dot{C}HCH_2- \\ -CH_2\dot{C}HCH_2 \end{matrix} \longrightarrow \begin{matrix} -CH_2-CH-CH_2- \\ | \\ -CH_2-CH-CH_2 \end{matrix} \quad (9.23)$$

For this to happen in polyethylene the radicals must either be formed in pairs in the 'primary' process or be able to migrate through the solid until pairs of radicals come close enough together to react. In fact both processes probably occur. The formation of pairs of radicals may occur in spurs and can also readily be envisaged if C—H bond breakage gives a hot hydrogen atom which abstracts a hydrogen atom from a neighbouring chain. Another possibility, although unsupported by experimental evidence, is that a positive ion on the main chain may take a hydrogen atom from a neighbouring chain giving a radical, or donate a proton to it, becoming a radical itself (indistinguishable effects). If this were to happen it would then be followed by neutralization of the positive ion by an electron, giving H_2 and a second radical. Radical migration through the solid may occur by a series of hydrogen atom abstractions either along a single chain or between molecules:

$$\begin{array}{ccc} -CH_2-\underset{\bullet}{C}H-CH_2- & & -CH_2-\underset{\displaystyle H}{\underset{|}{C}H}-CH_2- \\ & \longrightarrow & \\ -CH_2-\overset{\displaystyle H}{\overset{|}{C}H}-CH_2- & & -CH_2-\dot{C}H-CH_2- \end{array} \qquad (9.24)$$

It has been found that when D_2 is admitted to irradiated polyethylene, considerable H—D exchange takes place through reaction of polymer radicals with deuterium [33]:

$$-CH_2-\underset{\bullet}{C}H-CH_2- + D_2 \longrightarrow -CH_2-\underset{\displaystyle D}{\underset{|}{C}H}-CH_2- + D \qquad (9.25)$$

$$D + -CH_2-CH_2-CH_2- \longrightarrow HD + -CH_2-\underset{\bullet}{C}H-CH_2- \qquad (9.26)$$

This is surprising, since Reaction 9.25 would be expected to be forbidden on thermodynamic grounds. Probably the reaction is very slow. Occurrence of the reaction may assist radical migration to take place.

The main-chain unsaturation in irradiated polyethylene is probably formed partly by disproportionation as an alternative to Reaction 9.23 and partly by unimolecular loss of hydrogen from an excited polymer molecule. By analogy with low molecular weight compounds, several other processes must also occur in irradiated polyethylene and other polymers.

Effect of various substances on the chemical changes in irradiated polymers
The response of polymers to irradiation is markedly affected by the presence of other substances. Oxygen, as always, produces a powerful effect, which with polymers usually consists of a decrease in the tendency of the polymer to crosslink, and an increase in the tendency to degrade. Such an effect has been noted with polyethylene, polypropylene, polystyrene, poly(vinyl chloride), polyamides, polyacrylonitrile and polytetrafluoroethylene. At the same time as an alteration in the crosslinking pattern, oxygen causes peroxide, carbonyl, carboxylic and other groups to be formed. The amount of oxygen initially present in the material causes only a small effect, but large effects can occur if the sample is thin enough and the dose-rate low enough for oxygen to diffuse in during the course of an irradiation.

The deliberate incorporation of a variety of added substances into a polymeric material can markedly affect the change produced by irradiation. Work with poly(methyl methacrylate) provided an early example [34]. With this material it was found that the incorporation of 10 per cent of allyl thiourea, di-*m*-tolyl thiourea, aniline, 8-hydroxyquinoline or benzoquinone increased by up to about three times the dose needed to produce a given amount of degradation as measured by the decrease in viscosity of a solution of the polymer in chloroform. Such effects are important in considering the irradiation of commercial plastics, since these often contain additives including stabilizers, antioxidants, plasticizers, fillers and so on and will therefore not necessarily show the same behaviour as pure polymers. The effect with poly(methyl methacrylate) may be regarded as a protection of the polymer against radiation-induced degradation. Deliberate addition of suitable substances can help in the formulation of radiation-resistant materials. The effect also resembles the chemical protection which may be obtained in radiobiology. Some additives *increase* certain radiation effects both in polymers and in biological systems. In the case of polymers the mechanism may include transfer of energy of electronic excitation, positive charge and electron transfer, hydrogen atom transfer and many other processes.

Several investigators have irradiated polymers swollen with various solvents or in solution. It is generally found [35] that polymers which crosslink when irradiated in the solid state can become crosslinked when irradiated under these conditions. With solutions the effect is quite striking in appearance. The viscosity of the solution rises slowly with increasing dose, then at a certain critical dose the solution suddenly gels. Prolonged irradiation increases the crosslinking density so much that the gel can no longer take up all the solvent, and the gel then exudes solvent and breaks away from the container wall.

The usual relationship between the critical gel dose and the composition of the solution is shown formally in Fig. 9.4 [36]. With aqueous solutions the radicals formed from the water are very reactive, and can form radicals from the polymer which then join together to form crosslinks. Consequently the number of polymer molecules affected per 100 eV absorbed in the whole system can be of the same order of magnitude in a solution as in the irradiation of the pure polymer, and since the number of molecules to be affected is less, a higher proportion of molecules may be affected for a given dose. In the approximate concentration range 1–100 per cent, crosslinks form between different molecules and the critical gel dose decreases with increasing dilution. Below about one per cent however, the crosslinks tend to form intramolecularly and, even if there is some increase in molecular weight, the polymer never forms a network which extends through the system. With organic solvents the radicals generally tend to be relatively inactive and the predominant effect of radiation on the polymer is by direct action modified by 'energy transfer' effects. Accordingly the critical gel dose does not decrease as much on dilution as with aqueous solutions. Below around ten per cent of polymer, the polymer radicals tend to react with solvent radicals instead of with each other so that crosslinking is disfavoured. Direct action on the polymer still produces degradation however, so that the balance between degradation and crosslinking shifts in favour of degradation and gel does not form.

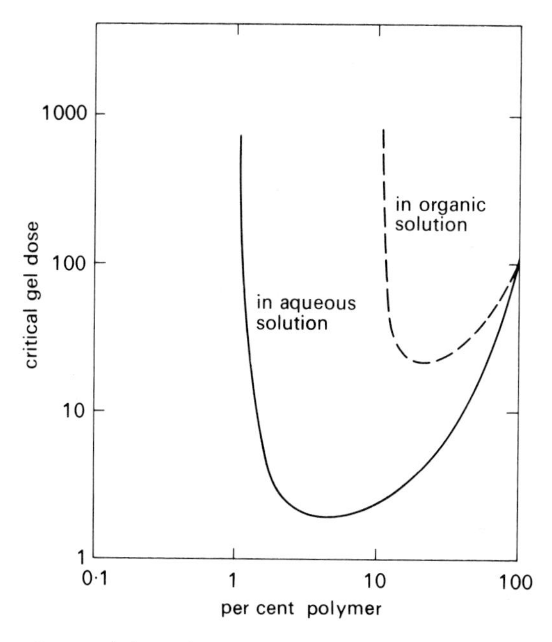

Fig. 9.4 Expected dependence of the minimum dose needed to form a gel on the percentage of polymer in solution. The two curves shown have similar shapes, but for different reasons, as explained in the text.

Special effects are observed when polymers are irradiated in the presence of vinyl monomers. These are discussed below (p. 225).

Physical changes in irradiated polymers The crosslinking or degradation of polymers by radiation causes striking changes in physical properties in general accordance with knowledge of polymer science. For example, low-density polyethylene melts at about 115°C but, after sufficient crosslinks have been introduced, it no longer converts into a liquid on heating above the normal melting point but is converted instead into a rubber-like material. This effect is shown in Fig. 9.5. The change in melting behaviour occurs quite suddenly, at the stage when a three-dimensional gel first extends through the system. Several other discontinuities also occur at the gel point.

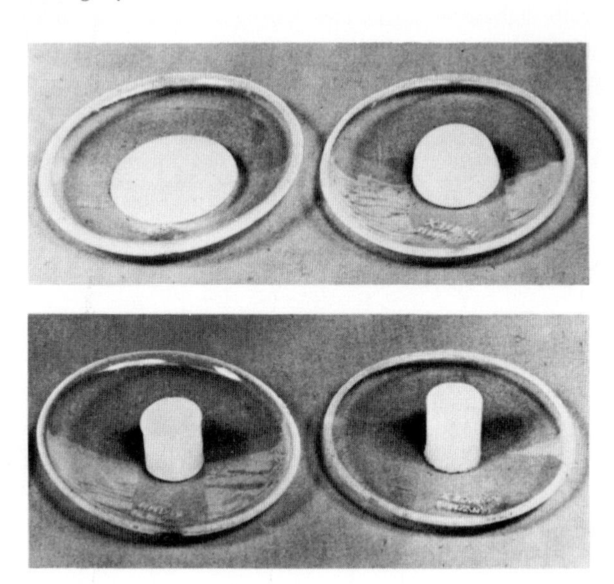

Fig. 9.5 Effect of heat on polyethylene. Identical samples had been heated for 1 hour at 180°C after irradiation to various doses.

The effect of radiation on the melting properties of polymer is the basis for a method which has been used commercially to produce shrinkable materials for packaging and other purposes. The first stage in the process is to irradiate the polymer beyond the gel point. The material is then heated to melt the crystalline regions in the polymer, and then expanded. While still under stress it is cooled so as to reform the crystals and so lock the material in the new shape. If the material is now placed round an object and then heated, the crystallites melt and the crosslinks tend to draw the material back to its original shape ('memory effect'), so forming a tight sheath round the object. An example of the application of shrinkable polyolefin is shown in Fig. 9.6.

The dose to reach the gel point may be calculated in the following way. Consider a molecule *A* which is already joined by a crosslink to another molecule and which may perhaps be linked to still further molecules. The average number of monomer units in *A* which are linked to

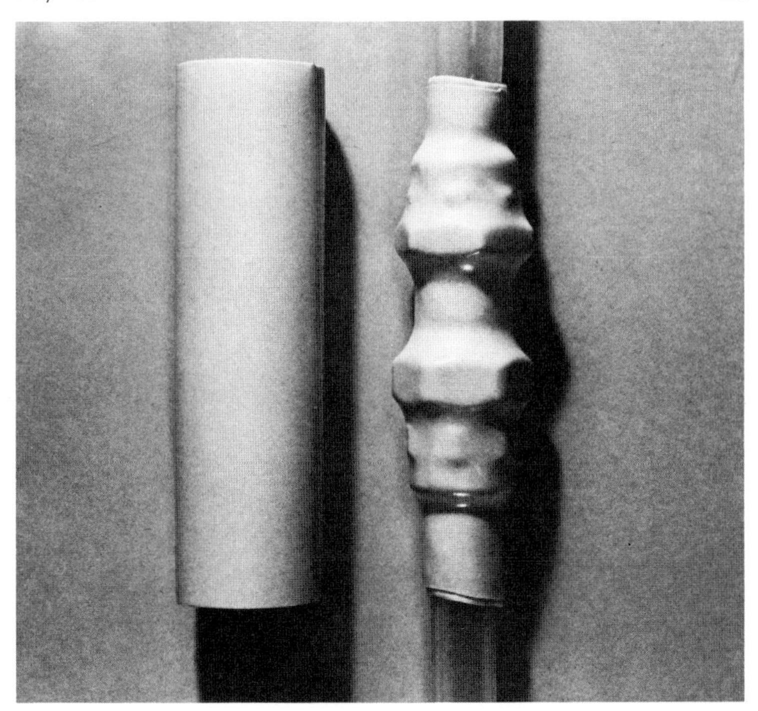

Fig. 9.6 Heat-shrinkable material (Raychem). The tubing on the left is made using the method described in the text. When it is placed round an object of irregular shape and then heated above 135°C it shrinks as shown on the right.

still further molecules, n, is equal to the probability p_c that any one monomer unit has a crosslink attached to it multiplied by the number of monomer units available for further crosslinking, that is, $P-1$, if P is the degree of polymerization of the molecule:

$$n = p_c(P-1) \tag{9.27}$$

If n is less than one, then the probability that an already crosslinked molecule will be linked to one more molecule will decrease steadily in progressing from one molecule to the next. Hence the system cannot possess an infinite network. On the other hand if n is greater than one, then an infinite network can arise. The critical condition for the gel point is where $n = 1$. At this point:

$$(p_c)_{crit}(P-1) = 1 \tag{9.28}$$

or since P is large:

$$(p_c)_{crit}P = 1 \tag{9.29}$$

that is, at the gel point, on the average, every molecule is crosslinked, or, since one crosslink joins two molecules, on the average there is one crosslink for every two molecules initially present. In practice all molecules do not have the same degree of polymerization, and the condition for gel formation in a sample of polymer containing molecules of different sizes is that there should be one monomer unit crosslinked per weight-average molecule originally present. If an average of G crosslinks are formed for each 100 eV absorbed during the course of the irradiation, and $\overline{M}_w$ is the weight-average molecular weight, then the dose in rads to reach the gel point is given by:

$$D_{gel} = \frac{6.023 \times 10^{23} \times 100}{2 \times 6.24 \times 10^{13} G\overline{M}_w} \tag{9.30}$$

Determination of the dose to gel can be used to measure G providing $\overline{M}_w$ is known for the sample.

Above the gel point, that part of the sample which forms an infinite network is insoluble in solvents which would have dissolved the original polymer. This part is called the gel fraction. The other molecules in the sample are still soluble, even though some of them have become crosslinked. This part is called the sol fraction. For molecules which have an initially random distribution of molecular weights, the following relationship [37] holds:

$$S + \sqrt{S} = p_0/q_0 + 1/q_0\overline{P}_n D \tag{9.31}$$

where S is the sol fraction, p_0 the proportion of monomer units fractured per unit radiation dose, q_0 the proportion of monomer units crosslinked per unit radiation dose, $\overline{P}_n$ the number-average degree of polymerization, and D the dose. The relationship is not strictly valid for polymers which do not initially possess a random distribution of molecular weights but, even for such a polymer, it is obeyed after large radiation doses. A plot of $S + \sqrt{S}$ against the reciprocal of the dose can be used to measure the G-values for crosslinking and degradation although, since the values obtained are for large doses, they may not be representative of the initial G-values. A plot of some typical experimental results for polyethylene is shown in Fig. 9.7.

Although a crosslinked polymer is no longer totally soluble beyond the gel point, it will still swell in solvents which would have dissolved the original polymer. At doses well above the gel point the amount of swelling decreases steadily as the density of crosslinks increases. G-values for crosslinking can be calculated from swelling measurements.

The continued introduction of crosslinks into an amorphous polymer leads to an increase in elastic modulus. For a three-dimensional network in an amorphous polymer the elastic modulus E is given by:

$$E = 3\rho RT/\overline{M}_c \qquad (9.32)$$

where ρ is the density of the polymer, R the gas constant, T the absolute temperature and $\overline{M}_c$ the number-average molecular weight between crosslinks. $\overline{M}_c$ is equal to $0.48 \times 10^6/GD$ where G is the number of crosslinks formed per 100 eV and D is the dose in megarads. The derivation of Equation 9.32 is subject to certain assumptions and simplifications, for which corrections can be applied [2]. After applying the corrections, elastic modulus can be used to measure the G-value for crosslinking.

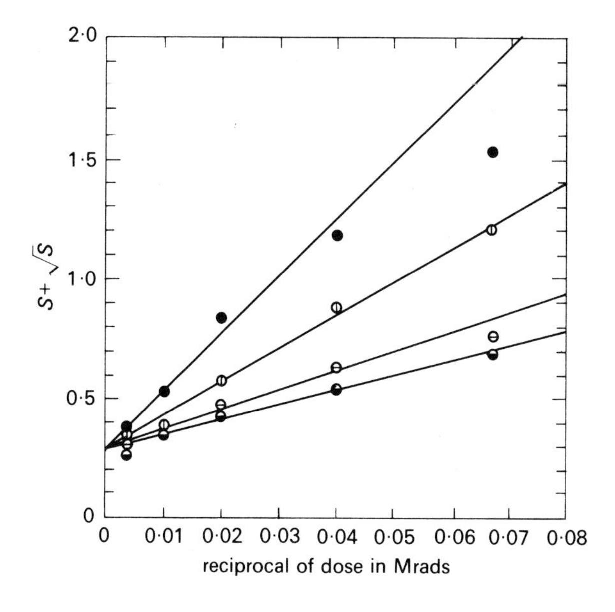

Fig. 9.7 Dependence of $S+\sqrt{S}$ on reciprocal of radiation dose in the irradiation of various samples of polyethylene. Such results can be used to calculate G-values for crosslinking and degradation providing the number average degree of polymerization is known.

Calculations can also be made of the relationship between any decrease in molecular weight produced by radiation and the viscosity of a solution of the polymer. Numerous experimental studies have been made of changes of physical properties of polymers on irradiation. In the case of commercial materials the measurement cannot always be related simply

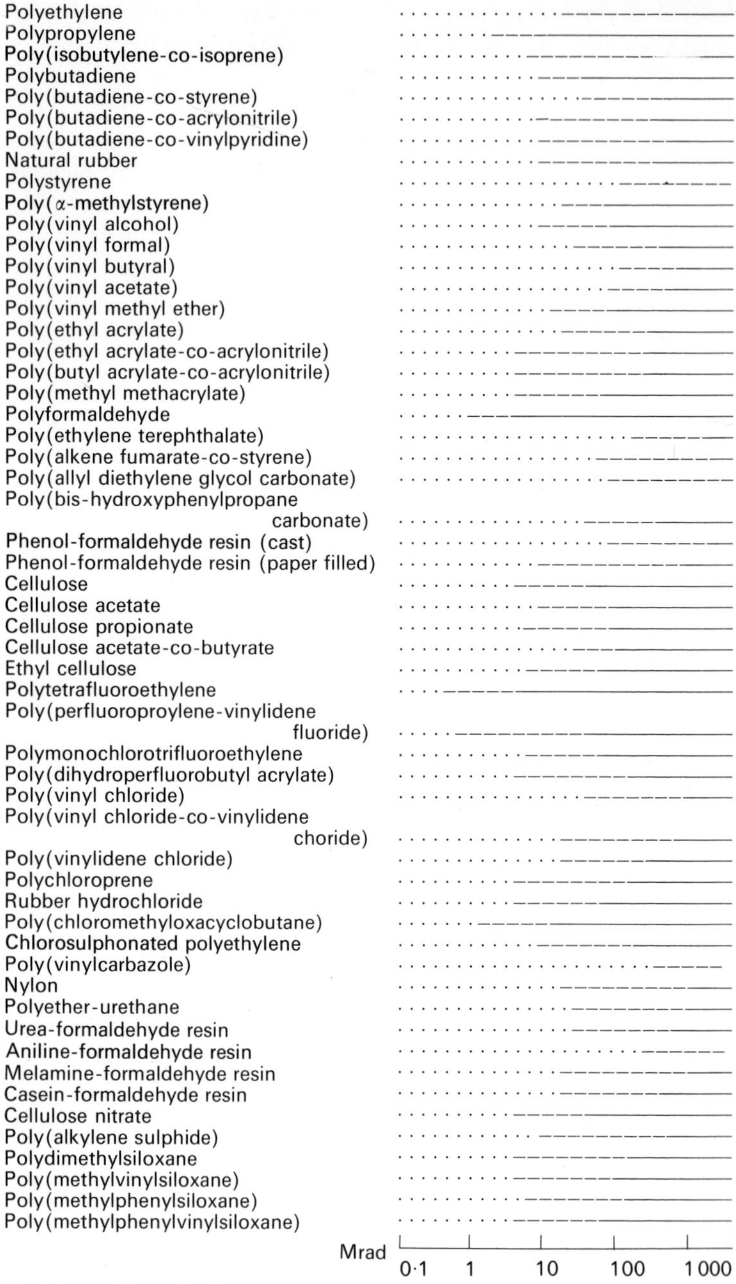

Polyethylene
Polypropylene
Poly(isobutylene-co-isoprene)
Polybutadiene
Poly(butadiene-co-styrene)
Poly(butadiene-co-acrylonitrile)
Poly(butadiene-co-vinylpyridine)
Natural rubber
Polystyrene
Poly(α-methylstyrene)
Poly(vinyl alcohol)
Poly(vinyl formal)
Poly(vinyl butyral)
Poly(vinyl acetate)
Poly(vinyl methyl ether)
Poly(ethyl acrylate)
Poly(ethyl acrylate-co-acrylonitrile)
Poly(butyl acrylate-co-acrylonitrile)
Poly(methyl methacrylate)
Polyformaldehyde
Poly(ethylene terephthalate)
Poly(alkene fumarate-co-styrene)
Poly(allyl diethylene glycol carbonate)
Poly(bis-hydroxyphenylpropane carbonate)
Phenol-formaldehyde resin (cast)
Phenol-formaldehyde resin (paper filled)
Cellulose
Cellulose acetate
Cellulose propionate
Cellulose acetate-co-butyrate
Ethyl cellulose
Polytetrafluoroethylene
Poly(perfluoroproylene-vinylidene fluoride)
Polymonochlorotrifluoroethylene
Poly(dihydroperfluorobutyl acrylate)
Poly(vinyl chloride)
Poly(vinyl chloride-co-vinylidene choride)
Poly(vinylidene chloride)
Polychloroprene
Rubber hydrochloride
Poly(chloromethyloxacyclobutane)
Chlorosulphonated polyethylene
Poly(vinylcarbazole)
Nylon
Polyether-urethane
Urea-formaldehyde resin
Aniline-formaldehyde resin
Melamine-formaldehyde resin
Casein-formaldehyde resin
Cellulose nitrate
Poly(alkylene sulphide)
Polydimethylsiloxane
Poly(methylvinylsiloxane)
Poly(methylphenylsiloxane)
Poly(methylphenylvinylsiloxane)

Mrad 0·1 1 10 100 1 000

Fig. 9.8 Effect of radiation on commercial plastics and elastomers.
.... negligible damage; – – – – – some damage; ——— severe damage

to the basic chemical changes, partly because of the presence of additives of various kinds as well as variable amounts of oxygen. However, the measurements are of practical importance as a guide to the performance of the materials in an irradiation field. Fig. 9.8 summarizes the results of many tests made on a variety of commercial materials [38].

A special kind of radiation effect which is quite different from those so far discussed is sometimes seen with natural rubber. If rubber is subjected to stress it is known to develop cracks on exposure to ozone. Such effects are sometimes seen in the neighbourhood of powerful sources of radiation where significant concentrations of ozone have been produced by the irradiation of the air.

Graft polymerization

If a dose of radiation is given to a gel formed by swelling a polymer with a monomer of a different type, then radicals formed from the polymer will initiate polymerization of the monomer, so that a graft copolymer will form (Fig. 9.9). Graft copolymers can also be formed by irradiating a

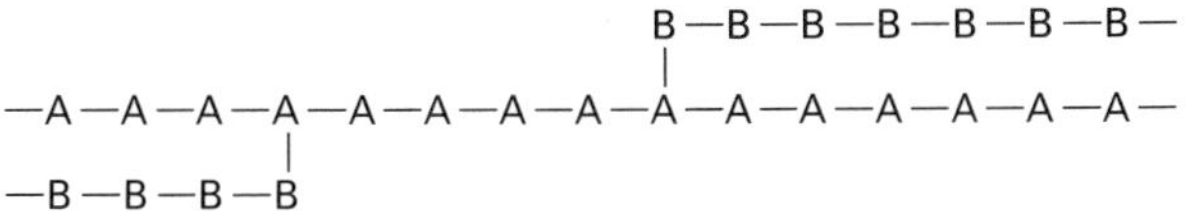

```
                                B—B—B—B—B—B—B—
                                |
—A—A—A—A—A—A—A—A—A—A—A—A—A—A—A—
         |
—B—B—B—B
```

Fig. 9.9 Graft copolymer.

polymer in the presence of oxygen to form peroxide and hydroperoxides, and then subsequently decomposing the peroxides (for example, by heating) while monomer is present. Both these methods necessarily produce some homopolymer together with the graft copolymer. A third method of making graft copolymers consists of irradiating the polymer to produce radicals, and then subsequently exposing it to monomer. This method gives less homopolymer. These methods have been used to prepare hundreds of interesting materials, some of them on a large scale. Examples include vinyl acetate grafted on cellulose acetate, styrene or acrylonitrile grafted on polyvinyl chloride, acrylic acid grafted on polythylene, and styrene grafted on cellulose. Publications on graft polymerization began to appear about 1955. At first it was thought that important industrial applications of the process would emerge, perhaps in the textile field, but despite a great deal of work, much of it reported in the open literature, progress in this direction has been slow. This is partly because the properties of the materials produced are not in general sufficiently valuable to justify the cost of development, and partly because there are in any case alternative methods of preparing such materials.

A combination of graft polymerization with crosslinking is encountered in the curing of unsaturated polyester resins. If a solution of an unsaturated polyester in a monomer such as styrene is irradiated, the unsaturated sites in the polyester become incorporated into the polymer formed from the monomer, so that a three-dimensional highly crosslinked material is formed by a chain reaction. The curing process has been conventionally carried out by blending a catalyst such as benzoyl peroxide into the uncured material and initiating the reaction by heat, but the radiation method offers certain practical advantages. The use of fast electrons to cure coatings (paints) incorporating unsaturated polyester resins or other materials is of considerable industrial interest.

Another process of industrial interest is the preparation of wood-plastic combinations by irradiation of wood which has been impregnated with monomer. Materials produced this way are not necessarily graft copolymers in the strict sense, but they can possess valuable properties which make them interesting from a commercial point of view.

PROBLEMS

1. If the yield of radicals from styrene is $G = 0.69$ per 100 eV, what would Equation 9.8 predict for the initial rate of polymerization of styrene at 25°C at a γ-ray dose-rate of 10 rads per second (k_t/k_p^2 at 25°C $= 24\,000$ $\mathrm{M}^{-1}\,\mathrm{s}^{-1}$, density of styrene $= 0.92$ g cm^{-3}).

[Ans: $4.6 \times 10^{-6}\,\mathrm{M\,s^{-1}}$]

2. Show that the rate of cationic polymerization under conditions where Reactions 9.18 and 9.20 are of comparable importance is given by the expression:

$$v_p = k_{pc} v_i [M] / \{(v_i k_t)^{1/2} + k_{tx}[X]\}$$

Assume the rate constant for the reaction of negative ions with X^+ is equal to that for Reaction 9.20.

3. Assuming the constants used for Fig. 9.2, what would be the initial rate of ionic polymerization of styrene under the same conditions as in question 1 at a concentration of water of $5 \times 10^{-9}\,\mathrm{M}$?

[Ans: $6.6 \times 10^{-4}\,\mathrm{M\,s^{-1}}$]

4. How much hydrogen chloride would be produced in the irradiation of one gram of poly(vinyl chloride) with a dose of 100 Mrads if the *G*-value for HCl production is 7.0?

[Ans: 0.725 m moles]

5. What dose would be required to render infusible (*a*) *n*-dodecane (*b*) a polymer of weight-average molecular weight 2×10^5, if in each case the number of crosslinks formed per 100 eV is 2.0.

[Ans: (*a*) 1.4×10^9 rads (*b*) 1.2×10^6 rads]

REFERENCES

1. F. A. Bovey. *The Effects of Ionizing Radiation on Natural and Synthetic High Polymers.* Interscience, New York, N.Y., 1958.
2. A. Charlesby. *Atomic Radiation and Polymers.* Pergamon Press, New York, N.Y., 1960.
3. A. Chapiro. *Radiation Chemistry of Polymeric Systems.* Interscience, New York, N.Y., 1962.
4. F. L. Hopwood and J. T. Phillips. *Nature*, **143**, 640 (1939), 'Polymerization of liquids by irradiation with neutrons and other rays'.
5. D. S. Ballantine, P. Colombo, A. Glines and B. Manowitz. *Chem. Eng. Progr. Symp. Ser.*, **50**, (11), 267–70 (1954), 'Gamma-ray-initiated polymerization of styrene and methyl methacrylate'.
6. S. Okamura and S. Futami. *Int. J. Appl. Radiat. Isotopes*, **8**, 46–51 (1960), 'Ionic polymerization of styrene in methylene dichloride by ionizing radiation'.
7. T. H. Bates, J. V. F. Best and T. F. Williams. *Nature*, **188**, 469–70 (1960), 'Radiation-induced ionic reactions: the retardation of the homopolymerizations of α-methyl styrene and β-pinene by water'.
8. W. H. Seitzer, R. H. Goeckermann and A. V. Tobolsky. *J. Am. Chem. Soc.*, **75**, 755–6 (1953), 'β-ray initiation of polymerization of styrene and methyl methacrylate'.
9. E. J. Lawton, W. T. Grubb and J. S. Balwit. *J. Polymer Sci.*, **19**, 455–8 (1956), 'A solid state polymerization initiated by high-energy electrons'.
10. W. H. T. Davison, S. H. Pinner and R. Worrall. *Chem. and Ind.*, 1274 (1957), 'Polymerization of isobutene with high energy radiation'.
11. H. Yamaoka, F. Williams and K. Hayashi. *Trans. Faraday Soc.*, **63**, 376–81 (1967), 'Radiation-induced polymerization of nitroethylene'.
12. F. Williams, in *Fundamental Processes in Radiation Chemistry*, ed. P. Ausloos. Interscience, New York, N.Y., 1968, pp. 515–98, 'Principles of radiation-induced polymerization'.
13. A. Chapiro and P. Wahl. *Compt. rend.*, **238**, 1803–5 (1954), 'Sur la relation vitesse-intensité dans la polymérisation du styrolène par les rayons γ'.
14. A. Fontijn and J. W. T. Spinks. *Can. J. Chem.*, **35**, 1384–413 (1957), 'Addition of *n*-butyl mercaptan to 1-pentene on irradiation with X-rays or gamma-rays'.

15. R. Bensasson and A. Prevot-Bernas. *J. chim. phys.*, **53**, 93–5 (1956), 'Polymérisation radiochimique de l'acrylonitrile en solution et distribution spatiale des centres actifs primaires'.
16. F. Williams, K. Hayashi, K. Ueno, K. Hayashi and S. Okamura. *Trans. Faraday Soc.*, **63**, 1501–11 (1967), 'Radiation-induced polymerization by free ions. Part 3. Rate constants for cationic polymerization'.
17. A. Chapiro, in *Actions Chimiques et Biologiques des Radiations*, dixième série, ed. M. Haïssinsky. Masson, Paris, 1966, pp. 187–312, 'Polymérisations en phase solide amorcées par les radiations'.
18. R. B. Mesrobian, P. Ander, D. S. Ballantine and G. J. Dienes. *J. Chem. Phys.*, **22**, 565–6 (1954), 'Gamma-ray polymerization of acrylamide in the solid state'.
19. M. Letort and A.-J. Richard. *J. chim. phys.*, **57**, 752–61 (1960), 'Mécanisme de formation du polyacétaldéhyde par fusion du cristal d'acétaldéhyde monomère'.
20. A. J. Restaino, R. B. Mesrobian, H. Morawetz, D. S. Ballantine, G. J. Dienes and D. J. Metz. *J. Am. Chem. Soc.*, **78**, 2939–43 (1956), 'γ-ray initiated polymerization of crystalline monomers'.
21. A. Charlesby. *Rept. Progr. Phys.*, **28**, 463–518 (1965), 'Solid-state polymerization induced by radiation'.
22. S. Okamura, K. Hayashi and Y. Kitanishi. *J. Polymer Sci.*, **58**, 925–53 (1962), 'Radiation-induced solid-state polymerization of ring compounds'.
23. J. F. Brown Jr. and D. M. White. *J. Am. Chem. Soc.*, **82**, 5671–8 (1960), 'Stereospecific polymerization in thiourea canal complexes'.
24. D. M. White. *J. Am. Chem. Soc.*, **82**, 5678–85 (1960), 'Stereospecific polymerization in urea canal complexes'.
25. A. Charlesby. *Proc. Roy. Soc.* (London), **A215**, 187–214 (1952), 'Cross-linking of polythene by pile radiation'.
26. Based on A. J. Swallow. *Radiation Chemistry of Organic Compounds*. Pergamon Press, New York, N.Y., 1960, p. 149.
27. A. A. Miller, E. J. Lawton and J. S. Balwit. *J. Polymer Sci.*, **14**, 503–4 (1954), 'Effect of chemical structure of vinyl polymers on crosslinking and degradation by ionizing radiation'.
28. M. Dole, D. C. Milner and T. F. Williams. *J. Am. Chem. Soc.*, **80**, 1580–8 (1958), 'Irradiation of Polyethylene. II. Kinetics of Unsaturation Effects'.
29. C. D. Bopp, W. W. Parkinson and O. Sisman, in *Radiation Effects on Organic Materials*, eds R. O. Bolt and J. G. Carroll. Academic Press, New York, N.Y., 1963, pp. 183–244, 'Plastics'.
30. M. Ross and A. Charlesby. *Atomics and Atomic Technol.*, **4**, 189–94 (1953), 'Effect of pile irradiation on polymethyl methacrylate ("Perspex")-I'.
31. H. Y. Kang, O. Saito and M. Dole. *J. Am. Chem. Soc.*, **89**, 1980–6 (1967), 'The radiation chemistry of polyethylene. IX. Temperature coefficient of cross-linking and other effects'.
32. A. Charlesby. *Advan. Chem. Ser.*, **66**, 1–21 (1967), 'Radiation mechanisms in polymers'.
33. M. Dole and F. Cracco. *J. Phys. Chem.*, **66**, 193–201 (1962), 'Radiation chemistry of polyethylene. V. Hydrogen isotope exchange studies'.
34. P. Alexander, A. Charlesby and M. Ross. *Proc. Roy. Soc. (London)*, **A223**, 392–404 (1954), 'The degradation of solid polymethylmethacrylate by ionizing radiation'.

35. A. Charlesby and P. Alexander. *J. chim. phys*, **52**, 699–708 (1955), 'Réticulation des polymères en solution aqueuse par les rayons gamma'.
36. A. Henglein and W. Schnabel, in *Current Topics in Radiation Research*, **2**, 1–67, (1966), 'Radiation chemistry of synthetic macromolecules in solution, LET effects'.
37. A. Charlesby and S. H. Pinner. *Proc. Roy. Soc. (London)*, **A249**, 367–86 (1959), 'Analysis of the solubility behavior of irradiated polyethylene and other polymers'.
38. S. H. Pinner. *Rept. Progr. Appl. Chem.*, **43**, 463–76 (1958), 'Irradiation of polymerisation products'.

Substances of biological interest

Radiation-chemical processes are to be found at their most intricate in the irradiation of biological materials which, at the molecular level, may be defined as highly organized systems of complex molecules in a predominantly aqueous environment. Radiation must act on such systems mainly by causing a more or less random series of excitations and ionizations along the tracks of the fast charged particles, after which a rapid series of reactions of the types discussed in earlier chapters must take place. The excited molecules, highly reactive ions, electrons and free radicals must disappear within a fraction of a second, giving new molecules to which the biochemical mechanisms of the systems respond. The importance of the subject derives from the need to understand the basic processes taking place in radiotherapy and also those processes responsible for the harmful effects associated with undesired exposure to nuclear and other radiations. The subject is also important in connection with the use of radiation to sterilize pharmaceutical products and to preserve food.

This chapter discusses the effects of radiation on some substances of biological interest when they are present in an environment which is sufficiently simple to enable the chemical changes to be understood. Attention is focused mainly on the low LET irradiation of fairly dilute aqueous solutions, where the principles of indirect action apply, as discussed in Chapter 7. Some discussion of direct action is also given.

Carbohydrates

The carbohydrates form an important group of naturally occurring organic compounds, and also provide the monomeric unit of the polysaccharides. Ribose and deoxyribose are essential parts of the nucleic acids. Apart from the intrinsic interest, it is important to understand the radiation chemistry of the carbohydrates as an aid to understanding the radiation chemistry of the high molecular weight materials. D-glucose, a representative carbohydrate, has been studied the most thoroughly.

Most of the work has been with simple aqueous solutions where the 'primary' process is the formation of radical and molecular products as in Equation 7.12 (p. 141). Hydrated electrons react slowly, if at all, with glucose, but hydroxyl radicals react at about the same rate as with simple alcohols. The reaction is probably the same: abstraction of hydrogen atoms from positions which are α- to a hydroxyl group. Hydrogen atoms react more slowly than hydroxyl radicals but would be expected to abstract similarly. The formation of the principal organic radicals may be represented as follows:

CH_2OH C O H H C CH(OH) OH H OH C C H OH → CH_2OH C O H H C •COH OH H OH C C H OH (10.1)

→ CH_2OH C O H H C CH(OH) OH OH C Ċ H OH (10.2)

→ •CHOH C O H H C CH(OH) OH H OH C C H OH (10.3)

Paper chromatography and radioactive tracer methods have been used to identify the final products obtained after irradiation in the absence or presence of oxygen. Some typical 'initial' yields, obtained from graphs showing the concentration of products to be linear with dose up to about 2×10^{20} eV cm^{-3}, are shown in Table 10.1. An acidic polymer is formed after very high doses in the absence of oxygen, but no polymer appears when oxygen is present during irradiation.

Table 10.1 *Initial yields in the γ-irradiation of aqueous solutions of glucose* *(5×10^{-2}M)* [1]

Product	*G-value* *Absence of oxygen*	*Presence of oxygen*
Glucose	−3.5	−3.5
Gluconic acid	0.35	0.4–0.5
Glucuronic acid	0	0.9
Glucosone	0.4	0
Two-carbon fragments	0.85	0.8
Erythrose	—	0.25
Three-carbon fragments	0.8	—

In the absence of oxygen, the gluconic acid and glucosone (2-oxo-D-arabinoaldohexose) are probably formed by abstraction of hydrogen atoms from the radicals formed in 10.1 and 10.2.

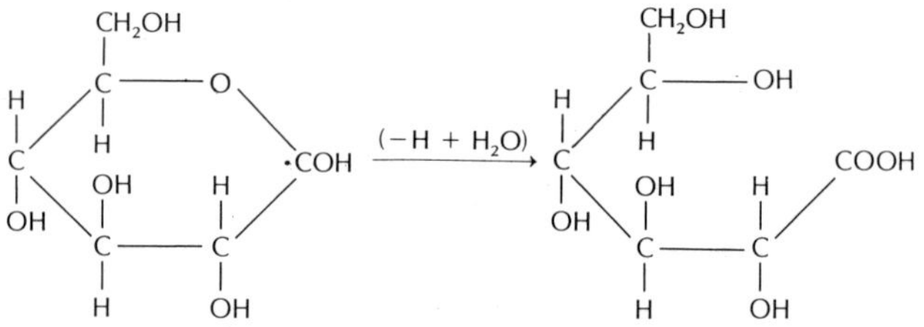

$$\xrightarrow{(-\mathrm{H}\,+\,\mathrm{H_2O})} \qquad (10.4)$$

$$\xrightarrow{(-\mathrm{H})} \qquad (10.5)$$

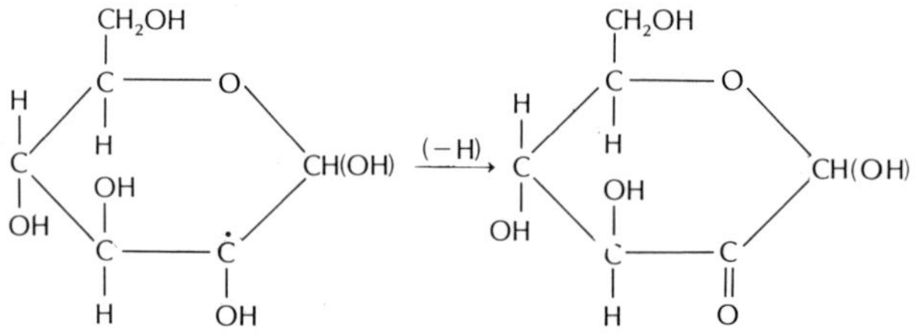

The radical abstracting the hydrogen atom will be one of the organic radicals formed in the system so that the reactions may be regarded as disproportionations. Carbon-carbon scission following loss of a hydrogen atom from a glucose radical could account for the two- and three- carbon fragments. Radicals would also be expected to dimerize: abstraction of a hydrogen atom from a dimer followed by addition of the dimer radical to another radical would yield a molecule of increased molecular weight, and repetition of the process would ultimately lead to a polymer. By a dose of 2×10^{20} eV cm^{-3} the concentration of a product formed with $G = 0.5$ would be 1.6×10^{-3}M. At such concentrations the products would be participating in further reactions such as reaction with hydrated electrons and organic radicals, so the full mechanism must be very complicated.

The formation of glucuronic acid on irradiation in the presence of oxygen may be due to reaction of oxygen with the radical formed by Reaction 10.3, yielding an unstable aldehyde which becomes oxidized to glucuronic acid. The absence of glucosone may be accounted for by the addition of oxygen to the radical formed in 10.2, followed by a chain of reactions including C—C scission, perhaps eventually giving arabinose and formaldehyde. The reaction of organic radicals with oxygen rather than with each other would account for the absence of polymer.

Less work has been done with glucose in the solid state, but it has been found that the overall yield for decomposition, *G*(-glucose), appears to be strongly dependent on the physical form. Yields as high as about 20 have been found for polycrystalline glucose.

Polysaccharides

Polysaccharides consist of carbohydrates joined together by ether linkages. The chemical changes occurring on irradiation are similar to those with simple carbohydrates like glucose. Reducing groups (carbonyl) and acid groups appear and carbon-carbon scission occurs. There is some evidence for dimerization with certain polysaccharides. The physical changes correspond to the chemical changes: in aqueous solution the viscosity decreases and in the 'dry' state degradation leads to an increase in solubility and a decrease in strength. Practical instances of radiation effects in polysaccharides are that paper becomes brittle, cellulose fibres and plastics lose strength, and fruits and vegetables become soft owing to degradation of the pectin component. These effects have to be taken into account in the use of radiation to sterilize certain medical supplies and in grafting to cellulose fibres [2].

Figure 9.8 (p. 224) includes an indication of the dose levels at which radiation damage becomes serious in cellulosic materials.

Amino acids and peptides

The radiolysis of amino acids and peptides has been studied principally as an aid to understanding the radiolysis of proteins. Hydrated electrons react with histidine, cysteine and cystine at neutral pH with rate constants about 10^{10} M^{-1} s^{-1}. They react with asparagine, arginine and the aromatic amino acids with k about 10^{8} M^{-1} s^{-1}. In all these cases the electrons are reacting with the side group. Electrons react with other amino acids with rate constants in the approximate range 10^{6}–10^{7} M^{-1} s^{-1}. All rate constants are larger in neutral solution than in alkaline solution. This is partly to be understood as a consequence of the availability of a loosely bound proton on the molecule in the neutral solutions. For example, the NH_2 group would become NH_3^+ which would be more reactive (pp. 150–151). The simple coulombic effect of charge is another factor (Equation 4.15, p. 79) [3]. Hydrated electrons react with peptides, containing the —CONH— group, many times faster than with the simple amino acids. For example, at pH 6.4 the rate constant for glycyl glycine is 2.5×10^{8} M^{-1} s^{-1} compared with 8×10^{6} $M^{-1} s^{-1}$ for glycine. This does not seem to be simply because the —CONH— group itself is reactive, since the rate constant for acetyl glycine is only 2×10^{7} M^{-1} s^{-1}. The pK_a for the amino group of the peptides is less than for the amino acids, so the greater reactivity may be associated at least partly with the presence of more loosely bound protons in the molecules [3].

Hydroxyl radicals react with amino acids at rates which were found in one of the early pulse radiolysis studies to vary from around 10^{7} M^{-1} s^{-1} to nearly 10^{10} M^{-1} s^{-1} depending on the amino acid and the pH [4]. Hydrogen atoms react with most aliphatic amino acids at rates which are hundreds of times less than for OH, but react with aromatic and sulphur-containing amino acids at comparable rates to OH.

Products detectable after X-radiolysis of a 1*M* aqueous solution of glycine are listed in Table 10.2. Aspects of the mechanism have been established with the aid of studies using added substances to compete for e^-_{aq} or OH [5]. The hydrated electrons may react with glycine according to the equations:

$$e^-_{aq} + NH_3^+CH_2COO^- \longrightarrow NH_3 + \dot{C}H_2COO^- \qquad (10.6)$$

$$\longrightarrow H + NH_2CH_2COO^- \qquad (10.7)$$

Table 10.2 *Initial yields in the X-irradiation of oxygen-free aqueous solutions of glycine (1M) [6]*

Product	*G-value*
NH_3	3.97
CHOCOOH	2.10
H_2	2.02
CH_3COOH	1.20
CO_2	0.90
HCHO	0.53
CH_3NH_2	0.19
HCOOH	0.085
O_2	<0.01

It is generally accepted that the hydroxyl radicals react according to:

$$OH + NH_3^+CH_2COO^- \longrightarrow H_2O + NH_3^+\dot{C}HCOO^- \qquad (10.8)$$

Hydrogen atoms from the 'primary' radiolysis of water and from Reaction 10.7 would be expected to react in a similar manner to OH radicals:

$$H + NH_3^+CH_2COO- \longrightarrow H_2 + NH_3^+\dot{C}HCOO- \qquad (10.9)$$

The organic radicals formed in Reaction 10.6 could perhaps react with glycine:

$$\dot{C}H_2COO^- + NH_3^+CH_2COO^- \longrightarrow CH_3COO^- + NH_3^+\dot{C}HCOO^- \qquad (10.10)$$

forming acetic acid, while the radicals from Reactions 10.8, 10.9 and 10.10 probably remove the hydrogen peroxide formed in Reaction 7.12 (p. 141) according to:

$$NH_3^+\dot{C}HCOO^- + H_2O_2 \longrightarrow NH_2^+{=}CHCOO^- + H_2O + OH \qquad (10.11)$$

The OH radicals formed in this reaction react according to Equation 10.8. Those extra organic radicals which do not react with hydrogen peroxide disproportionate giving the same imino acid as in Equation 10.11:

$$2NH_3^+\dot{C}HCOO^- \longrightarrow NH_2^+{=}CHCOO^- + NH_3^+CH_2COO^- \qquad (10.12)$$

The imino acid then hydrolyses:

$$NH_2^+{=}CHCOO^- + H_2O \longrightarrow NH_4^+ + CHOCOO^- \qquad (10.13)$$

$$\xrightarrow{+H^+} NH_4^+ + HCHO + CO_2 \qquad (10.14)$$

Small amounts of the dimeric products $(CH_2COOH)_2$, $(NH_2CHCOOH)_2$

and $NH_2CH(COOH)CH_2COOH$ have been found in irradiated solutions [7] and may originate in combination reactions of the organic radicals. Small amounts of some other products have been found too.

The yields of products from aqueous glycine are strongly dependent on pH. This must arise from changes in reactivity associated with the existence of various ionized forms of the organic free radicals and of the glycine itself, as well as from the expected changes in the reactions of the 'primary' species. Only limited information is available on the effect of oxygen on the yields. The ammonia yield appears to be reduced somewhat, less hydrogen and acetic acid are found and there is more hydrogen peroxide, glyoxylic acid and formaldehyde. Competition between Reactions 10.6, 10.7 and 10.9 and the formation of O_2^- (or HO_2) must play a part in the mechanism, while the $NH_3^+\dot{C}HCOO^-$ radicals may react with oxygen by electron transfer or addition, leading eventually to formation of one ammonia molecule and either one glyoxylic acid molecule or one formaldehyde molecule and one CO_2 per $NH_3^+\dot{C}HCOO^-$ [5].

Aqueous solutions of other amino acids and peptides undergo similar reactions to glycine on irradiation, but the other functional groups in the compounds are also attacked. Aromatic amino acids for example undergo reactions which are typical of aromatic compounds as well as reactions typical of simple amino acids. The absorption spectrum obtained immediately after OH addition to tryptophan is shown in Fig. 10.1 [8]. The three principal peaks are attributed to the addition of OH at three different positions in the indole ring. The absorption spectrum of the product

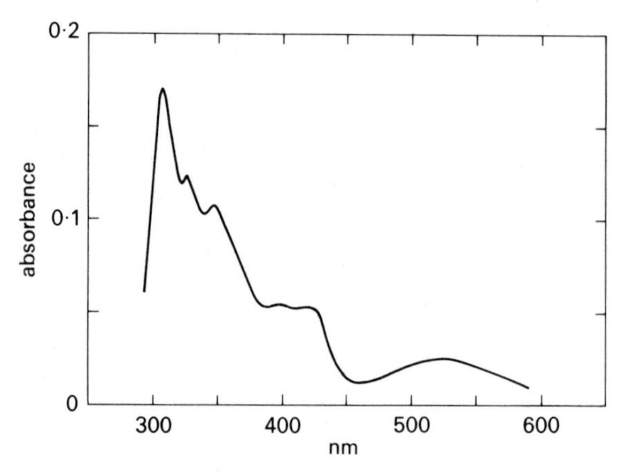

Fig. 10.1 Transient absorption spectrum of a nitrous-oxide saturated aqueous solution of tryptophan after a 2900 rad pulse of fast electrons. The spectrum is almost entirely attributable to species formed by reaction of the hydroxyl radical with tryptophan.

of the hydrated electron reaction exhibits an absorption in the region 300–450 nm, but the broad absorption at 500–550 nm is absent. Tryptophan is found to be consumed with $G = 1.7$ after fast electron irradiation of simple 2×10^{-2}M solutions at pH 7. Probably reduced and oxidized tryptophan radicals, represented as Trp—H• and Trp—OH• respectively, are reacting with each other in random disproportionations according to:

$$\text{Trp—H}\cdot + \text{Trp—OH}\cdot \longrightarrow 2\text{Trp} + H_2O \qquad (10.15)$$

$$\text{Trp—OH}\cdot + \text{Trp—OH}\cdot \longrightarrow \text{Trp} + \text{O—Trp} + H_2O \qquad (10.16)$$

$$\text{Trp—H}\cdot + \text{Trp—H}\cdot \longrightarrow \text{Trp} + H_2\text{—Trp} \qquad (10.17)$$

where O—Trp represents a stable hydroxylated product and H_2—Trp a stable reduced product. The rate constants for disappearance of the organic radicals are consistent with this view. On γ-radiolysis (that is, low dose-rate irradiation) of a 2×10^{-2}M solution, tryptophan is consumed only with $G = 0.7$. Probably at the lower dose-rate the tryptophan radicals react with the 'stable' irradiation products instead of with each other, so that tryptophan radicals revert to tryptophan, and molecules which differ more and more from the starting material are formed in small yield [8].

In the case of sulphur-containing amino acids the radiation chemistry is almost entirely that of the sulphur-containing groups (see next section). In the case of peptides an important response to radiation is degradation of the main chain, especially when oxygen is present [5].

Solid amino acids give rise to free radicals detectable by ESR, but the interpretation of the spectra is not always simple, especially as various free radicals appear to be present at various times after the irradiation. One of the earliest detailed studies was made on crystals of glycine [9]. in which the principal stable radical appears to be $NH_3^+\dot{C}HCOO^-$. Table 10.3 lists the products obtained after dissolving irradiated glycine in water.

Table 10.3 *Initial yields in the γ-irradiation of solid glycine* [10]

Product	*G-value*
NH_3	4.8
CHOCOOH	2.5
H_2	~0.2
CH_3COOH	2.3
CO_2	~0.2
HCHO	~0.03
CH_3NH_2	0.2

They are not very different from those obtained after irradiation of aqueous solutions except that very little hydrogen appears.

Thiols and disulphides

The —SH and —SS— groups are very susceptible to irradiation, and usually play a significant or even dominant role when any molecule containing them is irradiated. Their radiation chemistry is of special interest in view of the fact that organic sulphur compounds are good protective agents in radiobiology, since the mechanism of the protection may be chemical in nature.

Hydrated electrons react with both —SH and —SS— groups at rates which are close to diffusion-controlled, although for the thiols the rates are less when the solution is alkaline so that the group is in the form RS^-. Hydroxyl radicals and H atoms also react rapidly with both groups. The principal final irradiation products from aqueous solutions of thiols are the corresponding disulphides, together with the molecules corresponding to replacement of the —SH group by —H, hydrogen sulphide and hydrogen. Small amounts of other products may be present too. In the γ-irradiation of $10^{-2}M$ cysteine at pH 5–6, the G-values for cystine, alanine, H_2S and H_2 are 3.4, 2.6, 2.5 and 1.1 respectively [11]. With decreasing pH the yield of cystine remains about the same, while the alanine and H_2S decrease and the hydrogen increases. Oxygen can cause a manyfold increase in disulphide from thiols, owing to the development of chain reactions [12]. Irradiation of disulphides yields the trisulphides, the thiols, and RSO_2H and other oxygen-containing products. In the presence of oxygen the G-values of trisulphide and thiol are smaller and the G-values for oxygen-containing products are greater [13).

The reaction of the hydrated electrons with cysteine is principally to form hydrogen sulphide (in its anionic form SH^-) according to:

$$e^-_{aq} + RSH \longrightarrow R\cdot + SH^- \qquad (10.18)$$

although with other thiols hydrated electrons may also react to produce $RS\cdot$ and H_2. The immediate product of the reaction of hydrated electrons with disulphides is considered to be $RSSR^-$. This species can be observed by pulse radiolysis, and has a strong absorption in the visible as shown for cystamine in Fig. 10.2 [14]. The absorption maxima of these species are always in the region 400–450 nm, and extinction coefficients are about 8×10^3 M^{-1} cm^{-1}.

Hydroxyl radicals abstract hydrogen atoms from thiols:

$$OH + RSH \longrightarrow RS\cdot + H_2O \qquad (10.19)$$

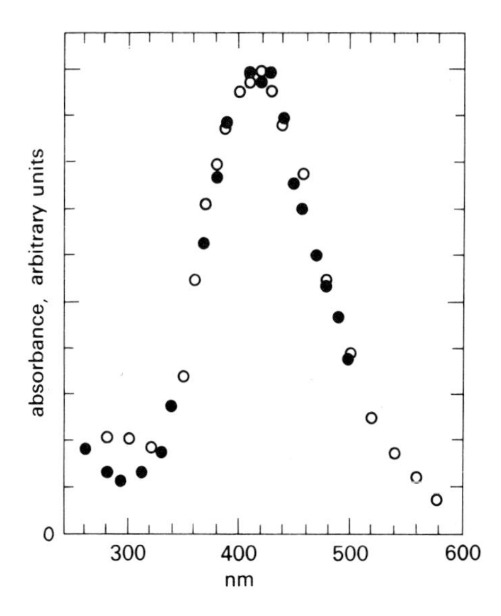

Fig. 10.2 Transient spectrum attributable to $RSSR^-$ produced by allowing hydrated electrons to react with cystamine (●). The same species can be formed by allowing OH to react with cysteamine (○).

The radical may react with the anionic form of the thiol according to:

$$RS\cdot + RS^- \rightleftharpoons RSSR^- \tag{10.20}$$

giving a transient species identical to that which can be made by electron addition to *RSSR*. The formation of this species from any given thiol is dependent on the concentration and pH of the solution, as shown for cysteamine in Fig. 10.3 [14]. Hydroxyl radicals are considered to react with disulphides to give an unstable molecule and a thiol radical [13]:

$$\mathrm{OH} + RSSR \longrightarrow RS\mathrm{OH} + RS\cdot \tag{10.21}$$

Hydrogen atoms react with thiols by two reactions:

$$\mathrm{H} + RS\mathrm{H} \longrightarrow \mathrm{H_2S} + R\cdot \tag{10.22}$$

$$\mathrm{H} + RS\mathrm{H} \longrightarrow \mathrm{H_2} + RS\cdot \tag{10.23}$$

They probably add to disulphides, giving *RSS*(H)*R*, the protonated form of $RSSR^-$, as an intermediate in the first instance.

It has long been accepted that owing to the low bond strength of S—H as compared with C—H, many organic free radicals are able to abstract hydrogen from thiols (cf. p. 207):

$$R\cdot + RSH \longrightarrow RH + RS\cdot \tag{10.24}$$

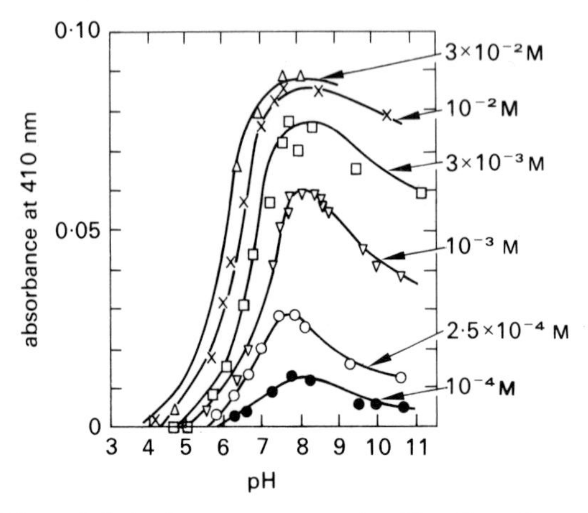

Fig. 10.3 Pulse radiolysis of aqueous cysteamine. The effect of concentration of cysteamine and pH can be understood as a consequence of the formation of $RSSR^-$ through reactions 10.19 and 10.20.

This is probably the normal fate of the radicals produced in 10.18 and 10.22. If an organic radical $X\cdot$ were to be formed in an irradiated system by loss of a hydrogen atom from a molecule XH, it could react with any thiol present by a reaction like 10.24, so that the molecule would be repaired. This reaction may be partly responsible for the protective action of sulphydryl compounds in some radiobiological systems [15].

Another established reaction in the radiolysis of sulphur compounds is the combination of $RS\cdot$ radicals, in the simple or complexed forms, to form disulphide, for example:

$$2RS\cdot \longrightarrow R_2S_2 \tag{10.25}$$

Such reactions account for the formation of disulphide in the radiolysis of oxygen-free solutions of thiols. Oxygen must enhance oxidation through the reaction of $RS\cdot$ or $RSSR^-$ to form organic peroxy radicals or O_2^- or HO_2. All or some of these radicals may then abstract hydrogen atoms from RSH, so leading to short chain reactions, for example:

$$RSSR^- + O_2 \longrightarrow RSSR + O_2^- \tag{10.26}$$

$$O_2^- + RSH \xrightarrow{+H^+} RS\cdot + H_2O_2 \tag{10.27}$$

$$RS\cdot + RS^- \longrightarrow RSSR^- \tag{10.28}$$

Many other reactions have been proposed for the radiolysis of thiols and disulphides in aqueous solution. Many of them are still speculative.

Various free radicals are detectable by ESR in sulphur compounds irradiated in the solid state. With both thiols and disulphides the radicals disappear with the formation of sulphur radicals. Cystine for example yields:

$$\begin{array}{c} HOOC{-}\underset{\displaystyle NH_2}{\underset{|}{CH}}{-}CH_2{-}S\cdot \end{array}$$

as the radical which is stable at room temperature [16]. Similar transfer reactions take place in the irradiation of mixtures of sulphur compounds with other compounds. With mixtures of cysteamine and DNA for example, DNA radicals appear to abstract hydrogen atoms from cysteamine so repairing the DNA and leading to the formation of $RS\cdot$ radicals from cysteamine [17]. This is similar to the reactions which occur in solution (cf. Equation 10.24) and lends further credence to the view that the process may have a part to play in radiobiological protection.

Proteins

Proteins consist of chains of different amino acid units joined together by peptide bonds. The long polypeptide chains are organized into a three-dimensional structure which is held together by secondary linkage including salt linkages between acidic and basic groups, hydrogen bonds, hydrophobic bonds and disulphide links. Hydrated electrons react with proteins with rate constants up to about 10^{11} M^{-1} s^{-1} [18]. For the unfolded molecule gelatin, the rate constant is approximately that expected from the amino acid composition and from the rate constants for the reaction of hydrated electrons with individual amino acids, of which histidine, cysteine, cystine, asparagine, arginine and the aromatic amino acids are the most important (see above, p. 234). For the folded proteins lysozyme and ribonuclease the rate constants are several times less than expected on the simple view. This is probably because hydrated electrons collide less frequently with protein molecules when they are in the compact form, and furthermore the reactive —SS— groups are shielded by unreactive amino acid units. Similar considerations must apply to the reactions of OH and H, although the reactions of these species cannot be studied directly.

Indications of the site of attack on proteins can also be gained from the transient absorption spectrum obtained after pulse radiolysis. The transient spectra observed in aqueous lysozyme solutions after a 0.2

microsecond electron pulse are shown in Fig. 10.4 [19]. The suppression of the peaks at 420 nm and 320 nm by the electron scavenger N_2O and the OH scavenger *t*-butanol respectively shows that the 420 nm peak must be due to a product formed from the hydrated electron, and the 320 nm peak must be due to a product formed from the OH radical. The 420 nm peak resembles that observed after reaction of the hydrated electron with simple disulphides (Fig. 10.2) and the 320 nm peak resembles that observed after reaction of OH with tryptophan (Fig. 10.1), so cystine and tryptophan may be the principal sites of attack giving rise to the absorptions observed. In the acid solutions containing *t*-butanol, H atom attack may be taking place at various aromatic sites.

Although the results of Fig. 10.4 indicate location of electrons on cystine residues, and OH radicals on tryptophan, reaction at other sites

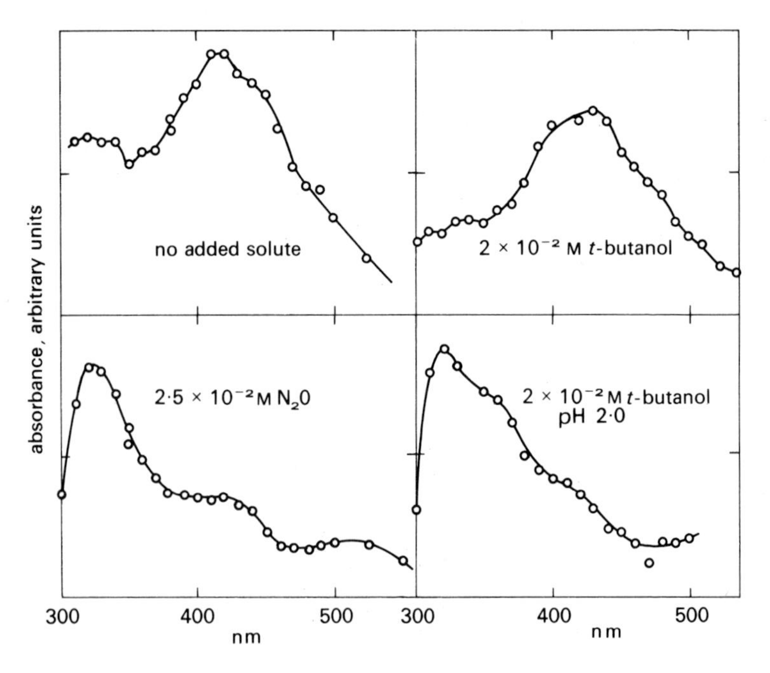

Fig. 10.4 Transient absorption spectra seen in pulse radiolysis of aqueous lysozyme containing various added solutes. The resemblance of the spectrum in the presence of the OH scavenger *t*-butanol to that shown in Fig. 10.2 suggests that the 420 nm band is due to an electron adduct to disulphide. The spectrum in the presence of the electron scavenger N_2O resembles Fig. 10.1, suggesting that the 320 nm band is due to an OH adduct at tryptophan. In the presence of *t*-butanol at pH 2 the enzyme is being attacked by H, perhaps through addition at aromatic sites.

is not excluded and in fact, from other work, at least half the electrons and three quarters of the OH radicals may have reacted at other sites. The transients produced may simply not absorb strongly in regions which can be observed. Initial attack must be followed by a train of reactions. In the case of H addition to ribonuclease the first reaction is ascribed in part to an attachment of H atoms to —SS— links. This is followed by intramolecular reactions such as transfer of H atoms to aromatic groups, with k around $10^3\ s^{-1}$. The radicals so formed then disappear in intramolecular reactions with k around $1\ s^{-1}$ [20].

Information about the final chemical change in irradiated proteins has been sought by analysing the amino acids in hydrolysates of the material. Table 10.4 shows a typical result, obtained after giving an X-ray dose of 50 000 r to a 10^{-5}M solution of serum albumin. The dose employed in this experiment would have been sufficient to change about ten amino acid units in every protein molecule. The main conclusion to be drawn is that reaction takes place at a number of different amino acid sites. Measurements of the final absorption spectrum also give indications of sites of attack. In one study it was found that where proteins contained more tyrosine units than tryptophan, the optical density of the irradiated solution at around 280 nm showed an increase like that observed on

Table 10.4 *Change in amino acid content of serum albumin on X-irradiation in aqueous solution (10^{-5}M, 50 000 r)* [21]

Amino acid	*Initial content (grams per 100 g protein)*	*Final content (grams per 100 g protein)*
Glycine	1.88	1.72
Alanine	6.78	5.13
Valine	5.53	5.54
Leucine	11.90	10.75
Isoleucine	2.59	2.51
Proline	5.12	4.72
Phenylalanine	6.37	5.89
Arginine	6.01	6.30
Histidine	4.01	4.14
Lysine	12.8	10.75
Aspartic acid	10.6	10.8
Glutamic acid	17.4	13.1
Serine	4.42	4.55
Threonine	5.94	5.11
Tyrosine	5.15	4.51

irradiation of aqueous tyrosine itself but, when there were more tryptophan units than tyrosine, the optical density at 280 nm decreased, as with aqueous tryptophan [21].

The chemical work shows that hydrated electrons, hydroxyl radicals and hydrogen atoms attack proteins at a number of sites, producing chemical changes which must be essentially similar to those occurring in simple amino acids and peptides. The changes will affect all types of secondary linkage, so that an alteration of the conformation of the protein molecules is to be expected. Changes of conformation, as well as other physico-chemical changes resulting from the chemical attack, show up in measurements of sedimentation coefficient, viscosity, light scattering, solubility, chromatographic behaviour, electrophoresis, ion-binding and other physico-chemical properties. In one such study, using aqueous solutions of ribonuclease, it was found that irradiation in the presence of oxygen produced two fractions which differed from the original when examined by gel chromatography. From ultracentrifuge analysis one of these had the same molecular weight as the original and the other had twice the molecular weight. The one with the same molecular weight was ascribed to a denatured form of the original molecule, that is, to a form in which chemical alterations to the amino acid and peptide groups had led to a change to a more open conformation. The fraction with the doubled molecular weight was ascribed to a dimer of the original molecule (crosslinked, cf. pp. 212–213). All fractions, including the one with the same chromatographic behaviour as the original, contained some altered amino acid units as shown by amino acid analysis. In the absence of oxygen the denatured form of the original molecule did not appear as a separable component. Probably it is necessary for a protein radical to react with oxygen to produce this form, and when no oxygen is present the radicals dimerize instead. The dimeric form produced in the presence of oxygen must be produced by a route other than simple combination of two carbon radicals [22].

The chemical and physico-chemical changes produced by attack of hydrated electrons, hydroxyl radicals and hydrogen atoms must produce changes in enzyme activity, but not necessarily very effectively, since not every part of the molecule is equally essential to enzyme activity. Indeed it may be possible to alter certain amino acid units without producing any change in activity at all. Nevertheless several pieces of evidence show that enzyme activity is in fact lost on irradiation and that, on irradiation of pure enzymes in solution, indirect action via e^-_{aq}, OH or H generally predominates greatly over direct excitation or ionization of the protein molecule itself. First, as shown for carboxypeptidase [23] and many other pure enzymes, appreciable activity is lost on irradiation, the actual amount of enzyme inactivated per unit volume of solution being approximately constant for a given dose of radiation, independent of concentration. Hence the *proportion* inactivated is greatest at the lowest concentrations. If inactivation were due to direct action, the proportion inactivated would have been independent of concentration. Furthermore, numerous added substances will protect enzymes against the effect of irradiation, presumably mainly by competing for e^-_{aq}, OH or H. In particular, impure preparations of enzymes are much less sensitive than pure preparations. The logarithm of the amount of activity lost on irradiation is often found to be proportional to the dose, and this can be understood if the active and inactivated molecules are competing for the e^-_{aq}, OH or H on approximately equal terms.

If the simple assumption could be made that every enzyme molecule is either completely active or completely inactive, the yield of inactivation could be expressed as a *G*-value: when expressed in this way yields of up to $G \sim 2$ molecules inactivated per 100 eV absorbed in the solution are found for pure enzymes. However, the assumption is probably oversimplified since varying degrees of inactivation are to be expected. It is also possible for molecules to be damaged in such a way that they retain enzyme activity but lose it on standing or on mild heat treatment or on reduction followed by reoxidation. Such an effect has been seen with pepsin [24] and with numerous other enzymes. Classical enzyme kinetics express the reaction velocity v of an enzyme-catalysed reaction in terms of a maximum velocity V and a Michaelis constant K_m. K_m is inversely proportional to the affinity of the enzyme for the substrate. For a substrate concentration $[S]$ the following expression applies:

$$\frac{1}{v} = \frac{K_m}{V} \cdot \frac{1}{[S]} + \frac{1}{V} \qquad (10.29)$$

There is no reason why V and K_m should both decrease equally. With deoxyribonuclease I, both V and K_m decrease on irradiation [25], but with other enzymes different behaviour is found.

Ultimately it is desirable to know to what extent each of the changes produced by e^-_{aq}, OH and H contributes to inactivation of an enzyme. The conclusion must depend strongly on the nature of the enzyme, and on the irradiation conditions such as pH, presence or absence of oxygen, organic substances and so on. Generally speaking a distinction can be made between specific changes at the active site, and overall conformational changes. A simple solution of the problem is not to be expected. Even for one enzyme, such as the relatively thoroughly studied ribonuclease [22], a number of different processes appears to be contributing at the same time.

Electron spin resonance has been used to examine the radicals produced by irradiating solid proteins. Irradiation at 77°K produces a mixture of radicals but on warming to room temperature the pattern simplifies and most proteins exhibit only two main components, one due to an $RS\cdot$ radical, like that observed with simple sulphur compounds (p. 241) and one which could most simply be attributed to a radical of formula:

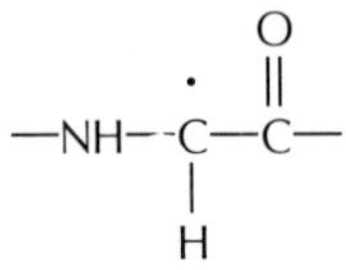

$$-NH-\overset{\cdot}{\underset{H}{\underset{|}{C}}}-\overset{O}{\overset{\|}{C}}-$$

The same two radicals are seen if the protein is irradiated at room temperature. It is most likely that a succession of transfer processes takes place within the solid, continuing until the stable radicals are formed [26]. When oxygen is present during irradiation, or when oxygen is admitted to the irradiated protein, a different ESR spectrum can be seen, attributable to $RO_2\cdot$ radicals formed by reaction of the carbon radicals with oxygen.

Although the stable carbon radicals in many irradiated proteins could, from the ESR evidence alone, be attributed to:

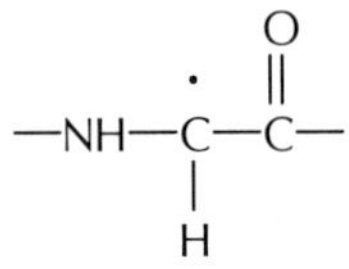

$$-NH-\overset{\cdot}{\underset{H}{\underset{|}{C}}}-\overset{O}{\overset{\|}{C}}-$$

experiments with tritiated hydrogen sulphide indicate that in fact a number of different kinds of radicals appears to be present. In these experiments TSH was admitted to proteins after irradiation to enable the reaction:

$$R\cdot + TSH \longrightarrow RT + \dot{S}H \qquad (10.30)$$

to take place. After labelling, the proteins were hydrolysed and the amino acids separated and counted. Only a little labelled glycine was found, suggesting that the simple formulation of the carbon radical must be incorrect. Moreover several other labelled amino acids were present [27]. Straightforward amino acid analysis on irradiated proteins and on fractionated irradiated proteins has shown loss of roughly similar amino acids, and the pattern is not very different from that in aqueous solution [22]. Evidently in the case of direct action on proteins, as well as in the case of indirect action, the observed changes are neither confined to single sites nor completely random.

Regardless of the precise mechanism, the chemical changes resulting from direct action on a protein must lead to loss of the biological functioning of the molecule. In a pioneering experiment, dried preparations of ribonuclease and myosin (muscle adenosinetriphosphatase) were exposed to large doses of X-rays (up to 10^8 roentgens). The enzyme activity was found to decrease exponentially with dose [28]. This can be explained mathematically on the target theory which, in its simplest original form, stated that a molecule becomes inactivated when an ionization occurs within it. On this view the activity of an enzyme at a dose *d* would be given by:

$$a_D = a_0 \exp(-d/D) \qquad (10.31)$$

where a_0 is the original activity and D is the dose to produce an average of one ionization per molecule. Now one roentgen corresponds to 1.61×10^{12} ionizations per gram of air and would correspond to 1.45×10^{12} ionizations per gram of protein if the energy required to produce an ion pair in protein were to be the same as in air. D would then be related to the molecular weight of the protein, M, by the expression:

$$D = \frac{6.023 \times 10^{23}}{1.45 \times 10^{12} M} \qquad (10.32)$$

Note the essential similarity between this expression and Equation 9.30 (p. 222). In the original experiment with ribonuclease, the dose required to reduce the enzyme activity to 37 per cent of its original value (that is, to e^{-1} of its original value) was $3{\cdot}4 \times 10^7$ roentgen. This quantity, often known as 'the D_{37} dose', is equal to D in Equations 10.31 and 10.32 and was used to give a value of 12 200 for the molecular weight of ribonuclease, in surprisingly good agreement with the subsequently established value of 13 680 considering the roughness of the assumptions made.

The elementary considerations of the simple target theory have been refined and extended and used to determine the molecular size of a large number of enzymes and other biologically active macromolecules [29]. During the course of the work it has become increasingly apparent that direct action can be modified by such factors as the presence of other substances (leading to a degree of protection or sensitization of the substance of interest) and the temperature of irradiation. Nevertheless it is an important truth about the action of radiation that the dose needed to produce a significant biological effect is inversely dependent on the molecular weight of the target material.

Electron transport systems

The flavins, nicotinamide adenine dinucleotide and its phosphate, ubiquinone and the cytochromes, all have the function in biological systems of accepting or donating reducing equivalents. Irradiation of these substances in systems whose composition is essentially similar to that of living cells is found to produce oxidations or reductions which in some ways resemble those which take place normally. The study of the reactions has not only thrown some light on how radiation may act on biological systems, but has also added to knowledge of the normal functions of the substances.

A simple model is provided by methylene blue, a reducible dyestuff with which hydrated electrons react at every collision. Consider the X-, γ- or fast-electron irradiation of a dilute solution containing methylene blue (about 10^{-5}–10^{-4} M) and an excess of an organic substance such as sodium formate, ethanol, sodium benzoate or sodium lactate (about 10^{-2}–10^{-1} M). In the absence of oxygen, irradiation is found to cause the methylene blue to become reduced to the colourless *leuco* form, while the added solute becomes simultaneously oxidized. After irradiation a slight recovery of the colour is often seen, probably because the molecular hydrogen peroxide is slowly oxidizing some of the *leuco* methylene blue back to the coloured form. On admitting air (oxygen) most of the original colour is restored. Similar effects take place when the methylene blue is present in a gel instead of a simple aqueous solution [30]. Many other dyestuffs show similar behaviour on irradiation, and effects in organic solvents and even in solid solution are not very different [31].

The mechanism of the reaction consists of the usual effect of radiation on water, followed by reaction of hydrated electrons with methylene blue to give a semiquinone free radical. At neutral pH the reaction of the hydrated electron may be represented as:

$$e^-_{aq} + Mb^+ \xrightarrow{+H^+} MbH\cdot^+ \qquad (10.33)$$

although the methylene blue and its semiquinone exist in different protonated forms at different pH values. The hydroxyl radicals and hydrogen atoms react with the organic substance present in excess giving an organic radical which, providing it is a stronger reducing agent than $MbH\cdot^+$, will react according to:

$$AH\cdot + Mb^+ \longrightarrow A + MbH\cdot^+ \qquad (10.34)$$

At acid pH hydrated electrons give hydrogen atoms which in turn give $AH\cdot$ so a similar reduction occurs. The semiquinone radicals formed in Reactions 10.33 and 10.34 can readily be observed by pulse radiolysis, and their absorption spectra, pK values and reactions specified. In strongly acid solutions (for example, 23N H_2SO_4) the semiquinone free radicals are stabilized by resonance, and remain as such [32], but in less acid, neutral or alkaline solutions the radicals disproportionate to yield the colourless *leuco* methylene blue and methylene blue itself:

$$2MbH\cdot^{+} \longrightarrow MbH_2^{+} + Mb^{+} \tag{10.35}$$

Leuco methylene blue is oxidized back to the dyestuff if air is admitted after irradiation. When oxygen is present during irradiation, methylene blue does not become reduced at all. This must be partly because of competition between oxygen and methylene blue for e^-_{aq} and $AH\cdot$ radicals. Also it is likely that methylene blue semiquinone radicals would, if formed, transfer electrons to oxygen, so reverting to the oxidized form.

In the absence of an excess of an organic compound to take them up, hydroxyl radicals appear to react with methylene blue like they do with any other aromatic compound. A mixture of products is formed.

Effects of radiation on redox systems of biological interest are in many ways similar to those on the methylene blue system. Riboflavin in particular responds in essentially the same way. Pulse radiolysis has been used to determine the properties of the semiquinone [33]. In the case of nicotinamide adenine dinucleotide (NAD) there is the special feature that the semiquinone does not disproportionate to yield the normal oxidized and reduced forms of the molecule, but dimerizes to yield a biologically inactive material [34]. With NAD the redox reaction has been carried out in the reverse direction by irradiating a nitrous oxide-saturated solution containing potassium bromide (about 10^{-1}M) and the reduced form of the coenzyme, NADH (about 10^{-4}M). In this system the hydrated electrons all react with N_2O to give OH radicals. If no bromide were present the OH radicals would react in various unspecific ways with NADH, but in the presence of bromide they are converted into a mild oxidizing agent which may be represented as Br_2^-, which then oxidizes NADH stoichiometrically:

$$OH + 2Br^- \longrightarrow Br_2^- + OH^- \tag{10.36}$$

$$Br_2^- + NADH \longrightarrow 2Br^- + NAD\cdot + H^+ \tag{10.37}$$

The NAD· radicals then dimerize. If oxygen is present, NAD· radicals react according to:

$$NAD\cdot + O_2 \longrightarrow NAD^+ + O_2^- \tag{10.38}$$

where NAD^+ represents the oxidized form of the molecule [35].

Some of the properties of the half-reduced form of ubiquinone, prepared by pulse radiolysis of a solution of ubiquinone in methanol, are shown in Table 10.5. In the case of cytochrome-c, it is found that certain free radicals which can be produced by radiation will specifically convert the oxidized form into the biologically active reduced form by transformation of the iron from ferric to ferrous, despite the existence of many other possible sites of reduction [37]. Hydroxyl radicals act on the molecule in a rather unspecific and complex way.

Table 10.5 *Some properties of ubisemiquinone* [36]

Neutral form	
Rate of formation by reaction of $\dot{C}H_2OH$ with ubiquinone	1.4×10^9 M^{-1} s^{-1}
Extinction coefficient at absorption maximum at 420 nm	3.0×10^3 $M^{-1}cm^{-1}$
Rate of disproportionation	4.8×10^7 M^{-1} s^{-1}
Rate of deprotonation to give anionic form	1.0×10^4 s^{-1}
p*K*	6.45
Anionic form	
Rate of formation by reaction of $\dot{C}H_2O^-$ with ubiquinone	2.0×10^9 M^{-1} s^{-1}
Rate of formation by reaction of solvated electron with ubiquinone	1.7×10^{10} M^{-1} s^{-1}
Extinction coefficient at absorption maximum at 445 nm	7.2×10^3 $M^{-1}cm^{-1}$

Nucleic acids and their components

Nucleic acids play a vital role in controlling living systems, and it is of the greatest importance for radiobiology to find out how they are altered by radiation. Most work has been done on deoxyribonucleic acid (DNA). This is a very high molecular weight material (molecular weight perhaps thousands of millions) with a main chain comprising alternate deoxyribose and phosphate units. Each deoxyribose unit is attached either to thymine or adenine or cytosine or guanine. Except during cell division, the nucleic acid chains usually exist as pairs of helices held together by hydrogen bonds between the bases (Fig. 10.5). In ribonucleic acids the deoxyribose units are replaced by ribose and the thymine by uracil. The conformations of the molecules are quite different from that of DNA.

Hydrated electrons do not react significantly with the deoxyribose or phosphate monomers of nucleic acids, but react with individual bases with rate constants in the region of 10^9–10^{10} M^{-1} s^{-1}. The rates are smaller in alkaline solutions than in neutral solutions probably partly because the electron distributions within the molecules are different owing to the existence of different tautomeric forms. The simple coulombic effect of charge would be another factor. In the polynucleotides there is a reduction in the rate constant (expressed per M of monomer unit) by more than an order of magnitude owing to a diminished collision frequency and, where base stacking occurs, an additional reduction of a few times owing to shielding of the base units from the solvent [38].

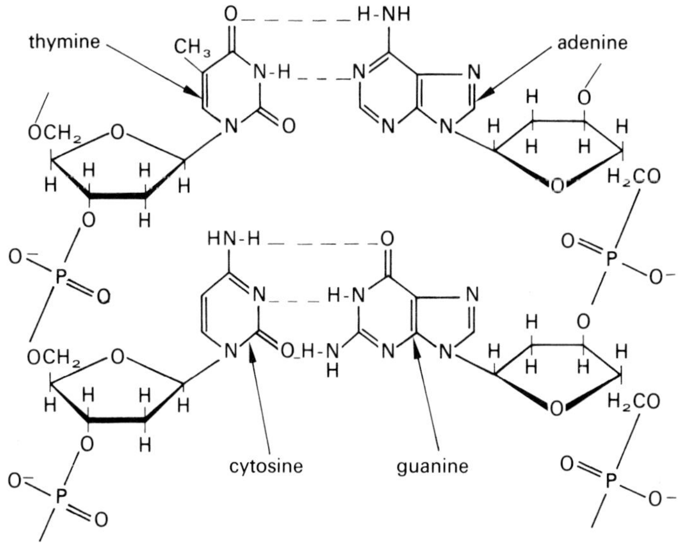

Fig. 10.5 Units of double stranded DNA molecule.

Hydroxyl radicals react with carbohydrates with rate constants which are only a few times smaller than their rate constants for reaction with purine and pyrimidine bases. In accordance with this, 20–25 per cent of OH radicals react with the deoxyribose units when aqueous solutions of DNA are irradiated. The remaining percentage reacts with the various bases on an approximately equal basis [39]. The rate constants must be influenced by collision frequency and base stacking in a similar way to the rate constants for the electron reaction. Hydrogen atoms appear to react principally by addition to the bases.

Model compounds have been used to study the chemical consequences of the attack of the primary species on nucleic acids. The radical produced by reaction of the hydrated electron with thymine appears to be capable of reverting to thymine (cf. tryptophan, p. 237) amongst other reactions but, when oxygen is present, it reacts with a rate constant of $8 \times 10^9\ M^{-1}\ s^{-1}$ to restore thymine with the concomitant production of hydrogen peroxide, according to the overall equation:

$$2T^- + O_2 + 2H_2O \longrightarrow 2T + H_2O_2 + 2OH^- \qquad (10.39)$$

This reaction may involve the intermediate formation of a peroxy radical of thymine [40]. The principal radical formed by reaction of OH radicals with thymine appears to be:

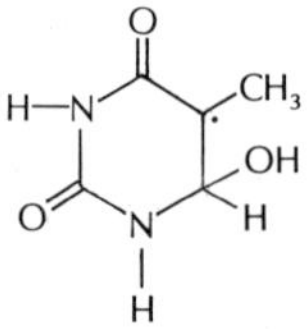

H—N, O, CH₃, OH, H, O, N, H

In the absence of oxygen this radical must take part in a large number of different reactions including reactions regenerating thymine but, when oxygen is present, it reacts to give a peroxy radical which then gives rise to a mixture of the *cis* and *trans* forms of the hydroperoxide with the —OOH group on the same carbon atom as the methyl group [41].

Some products detected after the X-irradiation of thymine in oxygen-free aqueous solution are listed in Table 10.6. Since e^-_{aq}, OH and H all

Table 10.6 *Initial yields in the γ-irradiation of aqueous solutions of thymine (10^{-2} M) [42]*

Product	*G-value*
Thymine	−2.5
Dihydrothymine	0.15
5-hydroxymethyl uracil	0.23
Thymine glycol: *trans*	0.42
Thymine glycol: *cis*	0.25
5(6)-hydroxyhydrothymine	0.38

react with thymine, the yield for loss of thymine would have been about $G = 6$ if it were not for the reactions regenerating the starting material. It is evident that a complete material balance has not yet been achieved. The radiation chemistry of cytosine and uracil is also complex, and that of the purines even more so [5, 39]. Relatively simple reactions take place however in the irradiation of cytosine or uracil in solutions containing Cu^{2+}. In this system the OH radicals add to the C=C bond of the cytosine or uracil and most of the radicals so formed react with the Cu^{2+} to form a glycol with a yield *G* about 2.3, only a little less than that of the primary yield of OH radicals. The rest of the hydroxyl adduct radicals react with Cu^{2+} by giving up a hydrogen atom to form isobarbituric acid derivatives (hydroxypyrimidines) [43].

In the irradiation of nucleotides (base-sugar-phosphate), important consequences of attack on the carbohydrate part of the molecule are the immediate liberation of free phosphate and the slow liberation of phosphate over a period of hours after irradiation. The slow liberation is speeded up in alkaline solutions [44]. The immediate liberation of phosphate is probably a consequence of oxidative attack like that on glucose (p. 231) at the carbon atom at which the phosphate is attached, and the slow liberation of phosphate is attributable to the formation of a labile phosphate ester by oxidation to carbonyl at one of the other carbon atoms. The effects are qualitatively the same whether oxygen is present during irradiation or not. Another consequence of attack on the sugar moiety is liberation of the free base from the nucleotide.

In the irradiation of nucleic acids themselves, the attack on the bases shows up as a chemical alteration of the base units with *G* about 0.2–0.6. Thymine is the most radiosensitive base. A nucleic acid hydroperoxide appears on irradiation in the presence of oxygen and is probably a hydroperoxide of the main chain thymine, since the hydroperoxides of the other bases are very unstable. Phosphate groups, analogous to those appearing in the radiolysis of nucleotide solutions, appear as end groups at the site of scission and are detectable using phosphomonoesterase [45]. Free bases appear, corresponding to those liberated from the nucleotides. It may be noted that all these effects are quite different from those obtained on ultraviolet illumination of DNA, where reversible hydration of the 5:6 double bonds of the pyrimidines takes place, as well as dimerization of adjacent pyrimidines, especially thymine.

The chemical changes in nucleic acids give rise to marked changes in physico-chemical properties. The earliest observation was that the viscosity of DNA solutions decreases on irradiation and then continues to decrease for many hours after the irradiation has stopped [46]. In principle the decrease could be due to elimination of interactions between particles, but studies with the ultracentrifuge, with light scattering and with enzymes have shown that a decrease in molecular weight (degradation) is responsible. It is most likely that the chemical process responsible for the degradation is the same as that which results in free phosphates being liberated from nucleotides, while the after effect is due, at least in part, to the slow hydrolysis of labile phosphate esters. The quantitative interpretation of the physico-chemical data is not simple, but most interpretations indicate that the *G*-value for production of single strand breaks in DNA is about $G = 0.5$–1.0. The molecular weight decreases when breaks occur sufficiently close to each other in the two strands. As well as degradation, there is also evidence that irradiation can produce some crosslinking.

Another important observation from titration studies, measurements of light absorption and so on, is that at the same time as the changes in molecular weight there is a marked breakage of the hydrogen bonds between the two components of the helices [47]. The breakage of hydrogen bonds at a single strand break may be visualized as in Fig. 10.6.

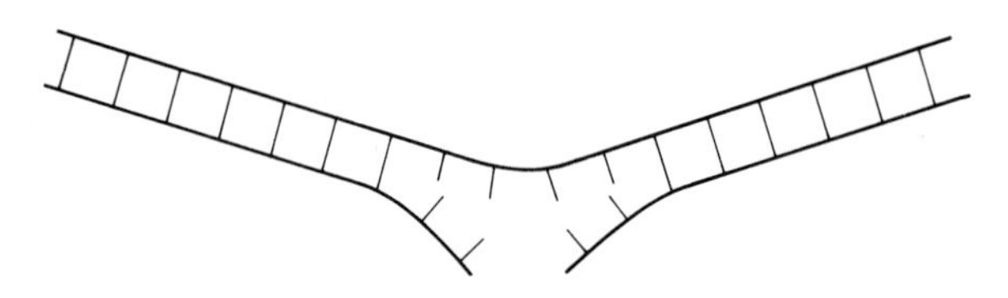

Fig. 10.6 Chain break in single strand of DNA accompanied by breakage of hydrogen bonds. Each of the lines between the strands corresponds to one base pair linked by two or three hydrogen bonds. The two strands are actually twisted to form a double helix.

Solid nucleic acids give rise to complex ESR spectra on irradiation. One component which can be identified with reasonable certainty in irradiated DNA is the adduct of a hydrogen atom to the thymine ring corresponding to the formula on p. 252 with OH replaced by H. This is identified by its characteristic 8-line spectrum, also shown by irradiated thymine itself. The chemical changes taking place in solid DNA have not been studied as extensively as those in aqueous solutions, but both single and double strand breaks have been observed. In contrast to those found in solution, the production of double breaks is linear with dose, showing that the double break is formed by a single primary event. There is also distinct evidence of crosslinking, especially when oxygen is absent [48]. Since the effects of radiation on solid DNA are by definition direct, very much larger doses are needed than in solution in order to show noticeable chemical or physical change, although the *G*-values are of the same order of magnitude in the two cases.

Lipids

The fatty components of cells have attracted much less attention than the water-soluble components. However, sterols have been irradiated in the pure state and in various solvents, and the products characterized. Despite the relative complexity of the molecules, attack seems to be mainly confined to certain reactive sites. For example, in the case of cholesterol, most of the reaction seems to occur in the neighbourhood of the double bond and the hydroxyl group. A table of the products isolated from irradiated sterols has been given elsewhere [49].

There has been considerable interest in the irradiation of fats in connection with the flavour changes in certain irradiated foods. In the absence of oxygen, fatty acid esters give acids, carbonyl compounds and other products. In the presence of oxygen, the radicals produced by radiation initiate an oxidation not unlike that occurring in autoxidation. Hydroperoxides are formed, and the final products also include acids and carbonyl compounds. Antioxidants are fairly effective in reducing the extent of the radiolysis.

Effects in organized biological systems

Owing to work such as that surveyed in this book and, especially, in this chapter, appreciable light has now been thrown on the types of molecular process which must occur when organized biological systems are irradiated with radiation of low LET. However, even in the best understood cases, knowledge is not yet either as qualitively complete or as quantitively precise as it might be. Further accurate research is therefore certainly needed, and is bound to throw up surprises. Not nearly enough is known about the chemical processes taking place in high LET irradiation, either of substances of biological interest or, as apparent from other chapters, of simpler substances. This is particularly unfortunate since high LET irradiation often produces much greater effects than low LET irradiation.

Radiation will continue to be used to treat cancer and to perform industrial processes on biological materials. Furthermore, as discussed in Chapter 1, there will be ever-increasing amounts of radiation in the environment arising from commercial applications and, especially, from nuclear power. It is therefore essential to continue to seek a deeper understanding of the chemical reactions taking place when biological substances are irradiated, particularly when the substances are as organized as they are in the living cell. An associated benefit of such work is that it will advance knowledge of normal biochemistry and biology.

As the full story of molecular radiation biology unfolds it will be found to differ in many respects from the story in this book. It will need to be told, in depth, later.

PROBLEMS

1. What would be expected from Equations 10.6–10.14 for the initial yields of the major products in the irradiation of an oxygen-free aqueous solution of glycine if the yields of 'primary' species are those given in Table 7.1 (p. 145) and the ratios of the rates of reactions 10.6:10.7 and 10.13:10.14 are taken to be 1.0 and 3.0 respectively?

[Ans: $G(NH_3) = 5.04$, $G(CHOCOOH) = 2.76$, $G(H_2) = 2.35$, $G(CH_3COOH) = 1.35$, $G(CO_2) = 0.92$, $G(HCHO) = 0.92$]

2. If the protonated form of the radical formed by reaction of a hydrated electron with acetone abstracts a hydrogen atom from cysteine to form an $RS\cdot$ radical, what G-values would be expected for (*a*) cystine (*b*) alanine (*c*) H_2S and (*d*) H_2 in the radiolysis of an aqueous solution containing equal amounts of acetone and cysteine buffered to pH 6, assuming the mechanism discussed on pp. 238–240, and taking the yields of 'primary' species from water to be those in Table 7.1 (p. 145), the hydrated electron rate constants to be those in Table 7.2 (p. 149) and the ratio of the rates of Reactions 10.23:10.22 to be 3.5. Neglect any slow reaction of hydrogen peroxide with cysteine.

[Ans: (*a*) 2.98 (*b*) 1.73 (*c*) 1.73 (*d*) 0.88]

3. A solution containing an enzyme and an excess of an organic compound which reacts with OH and H but not e^-_{aq} is given a single fast-electron pulse of 1.0×10^3 rads. What is the highest absorbance at about 425 nm which could be expected for an optical path length of 5 cm if all the hydrated electrons, produced with $G = 2.7$, go to disulphide groups giving $RSSR^-$ with an extinction coefficient of $9.0 \times 10^3 \text{M}^{-1}\ \text{cm}^{-1}$ if no other species absorbs at this wavelength?

[Ans: 0.125]

4. Plot a curve for the methylene blue concentration as a function of dose in rads for an oxygen-free solution containing 6×10^{-5}M methylene blue and an excess of an organic solute, on the assumption that the yield of primary radicals in the solution is $G = 5.8$.

[Ans: see Fig. 4 of reference 30]

5. What dose in rads would be required to produce an average of one single strand chain break per molecule in a sample of nucleic acid of molecular weight 5×10^6 (*a*) in 0.2 per cent solution if 0.5 breaks are produced per 100 eV absorbed in the solution (*b*) in the solid state if 0.5 breaks are also produced per 100 eV absorbed.

[Ans: (*a*) 770 (*b*) 3.9×10^5]

REFERENCES

1. G. O. Phillips. *Radiat. Res.*, **18**, 446–60 (1963), 'Chemical effects of ionizing radiations on aqueous solutions of aldohexoses'.
2. J. C. Arthur, Jr. in *Energetics and Mechanisms in Radiation Biology*, ed. G. O. Phillips. Academic Press, New York, N.Y., 1968, pp. 153–81, 'Radiation effects on cellulose'.
3. R. Braams, in *Pulse Radiolysis*, eds M. Ebert, J. P. Keene, A. J. Swallow and J. H. Baxendale. Academic Press, New York, N.Y., 1965, pp. 171–80, 'Reactions of the hydrated electron with amino acids, peptides and proteins'.
4. G. Scholes, P. Shaw, R. L. Willson and M. Ebert, in *Pulse Radiolysis*, eds M. Ebert, J. P. Keene, A. J. Swallow and J. H. Baxendale. Academic Press, New York, N.Y., 1965, pp. 151–64, 'Pulse radiolysis studies of aqueous solutions of nucleic acid and related substances'.
5. W. M. Garrison. *Current Topics in Radiation Research*, **4**, 43–94 (1968), 'Radiation chemistry of organo-nitrogen compounds'.
6. C. R. Maxwell, D. C. Peterson and N. E. Sharpless. *Radiat. Res.*, **1**, 530–45 (1954), 'The effect of ionizing radiation on amino acids. I. The effect of X-rays on aqueous solutions of glycine'.
7. B. M. Weeks and W. M. Garrison. *Radiat. Res.*, **9**, 291–304 (1958), 'Radiolysis of aqueous solutions of glycine'.
8. R. C. Armstrong and A. J. Swallow. *Radiat. Res.*, **40**, 563–79 (1969) 'Pulse- and gamma-radiolysis of aqueous solutions of tryptophan'.
9. D. K. Ghosh and D. H. Whiffen. *Mol. Phys.*, **2**, 285–300 (1959), 'The electron spin resonance spectrum of a γ-irradiated single crystal of glycine'.
10. G. Meshitsuka, K. Shindo, A. Minegishi, H. Suguro and Y. Shinozaki. *Bull. Chem. Soc. Japan*, **37**, 928–30 (1964) 'Radiolysis of solid glycine'.
11. V. G. Wilkening, M. Lal, M. Arends and D. A. Armstrong. *J. Phys. Chem.*, **72**, 185–90 (1968), 'The cobalt-60 γ radiolysis of cysteine in deaerated aqueous solutions at pH values between 5 and 6'.
12. J. E. Packer and R. V. Winchester. *Can. J. Chem.*, **48**, 417–21 (1970), '^{60}Co γ-radiolysis of oxygenated aqueous solutions of cysteine at pH 7'.
13. J. W. Purdie. *J. Am. Chem. Soc.*, **89**, 226–30 (1967), 'γ radiolysis of cystine in aqueous solution. Dose-rate effects and a proposed mechanism'.
14. G. E. Adams, G. S. McNaughton and B. D. Michael, in *The Chemistry of Ionization and Excitation*, eds G. R. A. Johnson and G. Scholes. Taylor and Francis, London, 1967, pp. 281–93, 'The pulse radiolysis of sulphur compounds Part I. Cysteamine and cystamine'.
15. P. Howard-Flanders. *Nature*, **186**, 485–7 (1960), 'Effect of oxygen on the radiosensitivity of bacteriophage in the presence of sulphydryl compounds'.
16. Y. Kurita and W. Gordy. *J. Chem. Phys.*, **34**, 282–8 (1961), 'Electron spin resonance in a gamma-irradiated single crystal of L-cystine dihydrochloride'.
17. M. G. Ormerod and P. Alexander. *Radiat. Res.*, **18**, 495–509 (1963), 'On the mechanism of radiation protection by cysteamine: an investigation by means of electron spin resonance'.
18. R. Braams. *Radiat. Res.*, **31**, 8–26 (1967), 'Rate constants of hydrated electron reactions with peptides and proteins'.
19. G. E. Adams, R. L. Willson, J. E. Aldrich and R. B. Cundall. *Int. J. Radiat. Biol.*, **16**, 333–42 (1969), 'On the mechanism of the radiation-induced inactivation of lysozyme in dilute aqueous solution'.

20. N. N. Lichtin, J. Ogdan and G. Stein. *Biochem. Biophys. Acta*, **263**, 14–30 (1972), 'Fast consecutive radical processes within the ribonuclease molecule in aqueous solution. I. Reaction with H atoms'.
21. E. S. Guzman Barron, J. Ambrose and P. Johnson. *Radiat. Res.*, **2**, 145–58 (1955), 'Studies on the mechanism of action of ionizing radiations. XIII. The effect of X-irradiation on some physico-chemical properties of amino acids and proteins'.
22. H. Dertinger and H. Jung. *Molecular Radiation Biology*. Springer-Verlag, New York, N.Y., 1970, pp. 115–33.
23. W. M. Dale. *Biochem. J.*, **34**, 1367–73 (1940), 'The effect of X-rays on enzymes'.
24. R. S. Anderson. *Brit. J. Radiol.*, **27**, 56–61 (1954), 'A delayed effect of X-rays on pepsin'.
25. S. Okada and G. Fletcher. *Radiat. Res.*, **16**, 646–52 (1962), 'Active site of deoxyribonuclease I. II. The mechanisms of the radiation-induced inactivation of deoxyribonuclease I in aqueous solution'.
26. T. Henriksen, in *Solid State Biophysics*, ed. S. J. Wyard. McGraw-Hill, New York, N.Y., 1969, pp. 201–41, 'The mechanisms for radiation damage and repair in solid biological systems as revealed by ESR spectroscopy'.
27. F. H. White Jr., P. Riesz and H. Kon. *Radiat. Res.*, **32**, 744–59 (1967), 'Free-radical distributions in several gamma-irradiated dry proteins as determined by the free-radical interceptor technique'.
28. D. Lea, K. M. Smith, B. Holmes and R. Markham. *Parasitology*, **36**, 110–18 (1944), 'Direct and indirect actions of radiation on viruses and enzymes'.
29. E. C. Pollard, W. R. Guild, F. Hutchinson and R. B. Setlow. *Progr. in Biophys. and Biophys. Chem.*, **5**, 72–108 (1955), 'The direct action of ionizing radiation on enzymes and antigens'.
30. M. J. Day and G. Stein. *Radiat. Res.*, **6**, 666–79 (1957), 'The action of ionizing radiations on aqueous solutions of methylene blue'
31. A. J. Swallow. *Radiation Chemistry of Organic Compounds*. Pergamon Press, New York, N.Y., 1960, pp. 175–85.
32. A. J. Swallow. *J. Chem. Soc.*, 1553–5 (1957), 'The preparation of stable free radicals in solution by means of ionising radiations'.
33. E. J. Land and A. J. Swallow. *Biochemistry*, **8**, 2117–25 (1969), 'One-electron reactions in biochemical systems as studied by pulse radiolysis. II. Riboflavin'.
34. G. Stein and A. J. Swallow. *J. Chem. Soc.*, 306–12 (1958), 'The reduction of diphosphopyridine nucleotide and some model compounds by X-rays'.
35. E. J. Land and A. J. Swallow. *Biochim. Biophys. Acta*, **234**, 34–42 (1971), 'One-electron reactions in biochemical systems as studied by pulse radiolysis. IV. Oxidation of dihydronicotinamide-adenine dinucleotide'.
36. E. J. Land and A. J. Swallow. *J. Biol. Chem.*, **245**, 1890–4 (1970), 'One-electron reactions in biochemical systems as studied by pulse radiolysis. III. Ubiquinone'.
37. L. K. Mee and G. Stein. *Biochem. J.*, **62**, 377–80 (1956), 'The reduction of cytochrome c by free radicals in irradiated solutions'.
38. P. C. Shragge, H. B. Michaels and J. W. Hunt. *Radiat. Res.*, **47**, 598–611 (1971), 'Factors affecting the rate of hydrated electron attack on polynucleotides'.
39. G. Scholes, in *Radiation Chemistry of Aqueous Systems*, ed. G. Stein. Weizmann Science Press, Jerusalem, 1968, pp. 259–85, 'Radiolysis of nucleic acids and their components in aqueous solutions'.

40. H. Loman and M. Ebert. *Int. J. Radiat. Biol.*, **18**, 369–79 (1970), 'The radiation chemistry of thymine in aqueous solution. Some reactions of the thymine-electron adduct'.
41. B. Ekert and R. Monier. *Nature*, **184**, B.A.58-B.A.59 (1959), 'Structure of thymine hydroperoxide produced by X-irradiation'.
42. M. N. Khattak and J. H. Green. *Int. J. Radiat. Biol.*, **11**, 577–82 (1966), 'Gamma-irradiation of nucleic-acid constituents in de-aerated aqueous solutions. III. Thymine and uracil'.
43. J. Holian and W. M. Garrison. *Nature*, **212**, 394–5 (1966), 'Radiation-induced oxidation of cytosine and uracil in aqueous solutions of copper(II)'.
44. M. Daniels, G. Scholes and J. Weiss. *J. Chem. Soc.*, 3771–9 (1956), 'Chemical action of ionising radiations in solutions. Part XVI. Formation of labile phosphate esters from purine and pyrimidine ribonucleotides by irradiation with X-rays in aqueous solution'.
45. G. Scholes. *Progr. in Biophys.*, **13**, 59–104 (1963), 'The radiation chemistry of aqueous solutions of nucleic acids and nucleoproteins'.
46. B. Taylor, J. P. Greenstein and A. Hollaender, *Cold Spring Harbor Symposia Quant. Biol.*, **12**, 237–46 (1947), 'The action of X-rays on thymus nucleic acid'.
47. R. A. Cox, W. G. Overend, A. R. Peacocke and S. Wilson. *Proc. Roy. Soc. (London)*, **B149**, 511–33 (1958), 'The action of γ-rays on sodium deoxyribonucleate in solution'.
48. J. T. Lett, K. A. Stacey and P. Alexander. *Radiat. Res.*, **14**, 349–62 (1961), 'Cross-linking of dry deoxyribonucleic acids by electrons'.
49. A. J. Swallow. *Radiation Chemistry of Organic Compounds*, Pergamon Press, New York, N.Y., 1960, pp. 188–95.

Appendix

Units and conversion factors

1 curie = 3.700×10^{10} disintegrations per second

1 kCi of cobalt-60 (point source) gives an exposure-rate of 1300 roentgens per hour at 1 m

1 kCi of caesium-137 (point source) gives an exposure-rate of 330 roentgens per hour at 1 m

Intensity of cobalt-60 radiation is reduced to one half by 1.06 cm of lead or 5.2 cm of concrete

Intensity of caesium-137 radiation is reduced to one half by 0.57 cm of lead or 3.8 cm of concrete

1 roentgen of X- or γ-radiation delivers 0.87 rad to air

1 roentgen of hard X- or γ-radiation delivers 0.97 rad to water

1 rad = 100 ergs per g
$= 10^{-5}$ J g^{-1}
$= 6.242 \times 10^{13}$ eV g^{-1}
$= 10^{-5}$ W s g^{-1}

1 eV g^{-1} = 1.602×10^{-14} rad

1 krad produces 3.1×10^{-6}M of a species formed with $G = 3$ in matter of unit density

1 eV = 1.602×10^{-19} J
$= 1.602 \times 10^{-12}$ ergs

1 calorie = 4.185 J

Energy of quantum in eV = 1.240×10^{3}/wavelength in nm

1 eV $\equiv 8.066 \times 10^{3}$ wave numbers (cm^{-1})
$\equiv 2.418 \times 10^{14}$ s^{-1}

1 Å = 10^{-1} nm = 10^{-4} μm

1 eV s^{-1} = 1.602×10^{-19} W

1 W = 6.242×10^{18} eV s^{-1}

1 eV per molecule = 23.06 kcal per mole
= 96.49 kJ per mole

1 kcal per mole = 4.185 kJ per mole

Charge on electron = 1.602×10^{-19} C
= 4.803×10^{-10} esu

Rest mass of electron = 9.109×10^{-28} g

Rest energy of electron = 0.511 MeV

Planck's constant = 6.626×10^{-34} J s
= 4.136×10^{-15} eV s

Velocity of light in vacuo = 2.998×10^{8} m s^{-1}

1 cm^3 per molecule per second = 6.02×10^{20} M^{-1} s^{-1}

1 M^{-1} s^{-1} = 1.66×10^{-21} cm^3 per molecule per second

Avogadro's number = 6.023×10^{23}

Volume of 1 mole of ideal gas at 273°K and 1 atmosphere pressure = 22.4 litres

Gas constant = 1.986 cal per mole per degree
= 8.314 J per mole per degree

1 atmosphere = 760 mm of mercury = 760 torr
= 1.003 kg per cm^2
= 1.013×10^{5} N m^{-2}
= 14.70 p.s.i.

1 bar = 10^{5} N m^{-2}

Index